UNDERSTANDING RESEARCH, SCIENCE AND TECHNOLOGY PARKS: GLOBAL BEST PRACTICES

Report of a Symposium

Committee on Comparative Innovation Policy:
Best Practice for the 21st Century

Board on Science, Technology, and Economic Policy

Policy and Global Affairs

Charles W. Wessner, Editor

NATIONAL RESEARCH COUNCIL
OF THE NATIONAL ACADEMIES

THE NATIONAL ACADEMIES PRESS
Washington, DC
www.nap.edu

THE NATIONAL ACADEMIES PRESS 500 Fifth Street, N.W. Washington, DC 20001

NOTICE: The project that is the subject of this report was approved by the Governing Board of the National Research Council, whose members are drawn from the councils of the National Academy of Sciences, the National Academy of Engineering, and the Institute of Medicine. The members of the committee responsible for the report were chosen for their special competences and with regard for appropriate balance.

This study was supported by: Contract/Grant No. SB1341-03-C-0032 between the National Academy of Sciences and the U.S. Department of Commerce; Contract/Grant No. OFED-381989 between the National Academy of Sciences and Sandia National Laboratories; Contract/Grant No. OFED-858931 between the National Academy of Sciences and Sandia National Laboratories; Contract/Grant No. SRS-0827103 between the National Academy of Sciences and the National Science Foundation; and Contract/Grant No. NAVY-N00014-05-G-0288, DO #2, between the National Academy of Sciences and the Office of Naval Research. This material is based upon work also supported by the Defense Advanced Research Projects Agency Defense Sciences Office, DARPA Order No. K885/00, Program Title: Materials Research and Development Studies, Issued by DARPA/CMD under Contract #MDA972-01-D-0001. Additional funding was provided by Intel Corporation, International Business Machines, M Square, the Association of University Research Parks, the Palo Alto Research Center, and Google. Any opinions, findings, conclusions, or recommendations expressed in this publication are those of the author(s) and do not necessarily reflect the views of the organizations or agencies that provided support for the project.

International Standard Book Number-13: 978-0-309-13789-8
International Standard Book Number-10: 0-309-13789-6

Limited copies are available from Board on Science, Technology, and Economic Policy, National Research Council, 500 Fifth Street, N.W., W547, Washington, DC 20001; 202-334-2200.

Additional copies of this report are available from the National Academies Press, 500 Fifth Street, N.W., Lockbox 285, Washington, DC 20055; (800) 624-6242 or (202) 334-3313 (in the Washington metropolitan area); Internet, http://www.nap.edu.

Printed in the United States of America

THE NATIONAL ACADEMIES

Advisers to the Nation on Science, Engineering, and Medicine

The **National Academy of Sciences** is a private, nonprofit, self-perpetuating society of distinguished scholars engaged in scientific and engineering research, dedicated to the furtherance of science and technology and to their use for the general welfare. Upon the authority of the charter granted to it by the Congress in 1863, the Academy has a mandate that requires it to advise the federal government on scientific and technical matters. Dr. Ralph J. Cicerone is president of the National Academy of Sciences.

The **National Academy of Engineering** was established in 1964, under the charter of the National Academy of Sciences, as a parallel organization of outstanding engineers. It is autonomous in its administration and in the selection of its members, sharing with the National Academy of Sciences the responsibility for advising the federal government. The National Academy of Engineering also sponsors engineering programs aimed at meeting national needs, encourages education and research, and recognizes the superior achievements of engineers. Dr. Charles M. Vest is president of the National Academy of Engineering.

The **Institute of Medicine** was established in 1970 by the National Academy of Sciences to secure the services of eminent members of appropriate professions in the examination of policy matters pertaining to the health of the public. The Institute acts under the responsibility given to the National Academy of Sciences by its congressional charter to be an adviser to the federal government and, upon its own initiative, to identify issues of medical care, research, and education. Dr. Harvey V. Fineberg is president of the Institute of Medicine.

The **National Research Council** was organized by the National Academy of Sciences in 1916 to associate the broad community of science and technology with the Academy's purposes of furthering knowledge and advising the federal government. Functioning in accordance with general policies determined by the Academy, the Council has become the principal operating agency of both the National Academy of Sciences and the National Academy of Engineering in providing services to the government, the public, and the scientific and engineering communities. The Council is administered jointly by both Academies and the Institute of Medicine. Dr. Ralph J. Cicerone and Dr. Charles M. Vest are chair and vice chair, respectively, of the National Research Council.

www.national-academies.org

Committee on Comparative Innovation Policy: Best Practice for the 21st Century*

Alan Wm. Wolff, *Chair*
Partner
Dewey & LeBoeuf LLP
and STEP Board

Kenneth S. Flamm, *Vice Chair*
Dean Rusk Chair in International Affairs
Lyndon B. Johnson School of Public Affairs
University of Texas at Austin
and STEP Board

Alice H. Amsden
Professor of Political Economy
Massachusetts Institute of Technology

Gail H. Cassell
Vice President, Scientific Affairs
and
Distinguished Lilly Research Scholar for Infectious Diseases
Eli Lilly and Company

Mary L. Good, *Vice Chair*
Donaghey University Professor
Dean, Donaghey College of Engineering and Information Technology
University of Arkansas at Little Rock
and STEP Board

Carl J. Dahlman
Henry R. Luce Associate Professor
Edmund A. Walsh School of Foreign Service
Georgetown University

Bronwyn Hall
Professor of Economics
University of California at Berkeley

Mark B. Myers
Senior Vice President, *retired*
Xerox

*As of December 2008.

Project Staff*

Charles W. Wessner
Study Director

Alan Anderson
Consultant

David E. Dierksheide
Program Officer

Sujai J. Shivakumar
Senior Program Officer

Adam H. Gertz
Program Associate

Jeffrey C. McCullough
Program Associate
(through August 2008)

*As of December 2008.

For the National Research Council (NRC), this project was overseen by the Board on Science, Technology and Economic Policy (STEP), a standing board of the NRC established by the National Academies of Sciences and Engineering and the Institute of Medicine in 1991. The mandate of the STEP Board is to integrate understanding of scientific, technological, and economic elements in the formulation of national policies to promote the economic well-being of the United States. A distinctive characteristic of STEP's approach is its frequent interactions with public- and private-sector decision makers. STEP bridges the disciplines of business management, engineering, economics, and the social sciences to bring diverse expertise to bear on pressing public policy questions. The members of the STEP Board* and the NRC staff are listed below:

continued on following page

*As of December 2008.

Joseph P. Newhouse
John D. MacArthur Professor of Health Policy and Management
Harvard Medical School

Arati Prabhakar
General Partner
U.S. Venture Partners

William J. Raduchel
Chairman
Opera Software ASA

Jack W. Schuler
Partner
Crabtree Partners

Alan Wm. Wolff
Partner
Dewey & LeBoeuf LLP

STEP Staff*

Stephen A. Merrill
Executive Director

David E. Dierksheide
Program Officer

Adam H. Gertz
Program Associate

Guruprasad Madhavan
Christine Mirzayan
Science & Technology Policy Fellow

Charles W. Wessner
Program Director

Jeffrey McCullough
Program Associate
(through August 2008)

Daniel Mullins
Program Associate

Sujai J. Shivakumar
Senior Program Officer

*As of December 2008.

Contents

PREFACE xiii

SUMMARY 1

I. INTRODUCTION 5

II. PROCEEDINGS

Welcome 41
Charles Wessner, National Research Council

Keynote Address I 44
Jeff Bingaman, United States Senate

Keynote Address II 47
C. D. Mote, Jr., University of Maryland

Panel I: Leading Asian Models of S&T Parks 53
Moderator: Lawrence Schuette, Office of Naval Research

China: Navigating at the Frontier of Life Sciences Silk Road 54
Zhu Shen, BioForesight

The Singapore Science and Technology Park 57
Yena Lim, Singapore Agency for Science, Technology and Research

Indian Science and Technology Parks 61
M. S. Ananth, Indian Institute of Technology-Madras

Discussant 66
Phillip H. Phan, Rensselaer Polytechnic Institute

Panel II: North American and European S&T Parks 70
Moderator: Peter Engardio, BusinessWeek

The English Experience 70
Jane Davies, Manchester Science Park, United Kingdom

Monterrey: International City of Knowledge Program 74
Jaime Parada, Research and Innovation Technology Park (PIIT)

Science and Technology Park Developments in Hungary 77
Ilona Vass, Hungarian National Office for Research and Technology

Initiatives in France 81
David Holden, Minatec

Keynote Address III 86
James Barker, Clemson University

Panel III: U.S. Parks: The Laboratory Model 92
Moderator: Kathryn Clay, U.S. Senate Energy Committee

U.S. and Global Best Practices: Sandia Science and Technology Park 93
Richard Stulen, Sandia National Laboratories

NASA Research Park 96
Simon (Pete) Worden, NASA Ames Research Center

The National Cancer Institute and NCI-Frederick 98
John Niederhuber, National Cancer Institute

Panel IV: U.S. Parks: University-Based Models 103
Moderator: Christina Gabriel, The Heinz Endowments

Research Triangle Park: Past Success and The Global Challenge 103
Rick L. Weddle, Research Triangle Park

Purdue Research Park 109
Victor Lechtenberg, Purdue University

Panel V: The Evaluation Challenge & Policy Synergies 113
Moderator: James Turner, U.S. House Science Committee

The Role of SBIR and State Awards 114
Robert McMahan, State of North Carolina

The Evaluation Challenge 117
Albert N. Link, University of North Carolina at Greensboro

Discussant 120
William Kittredge, U.S. Department of Commerce

Closing Remarks 123
Mary Good, University of Arkansas at Little Rock and Board on Science, Technology, and Economic Policy

III. RESEARCH PAPER

Research, Science, and Technology Parks: An Overview of the Academic Literature 127
Albert N. Link, University of North Carolina at Greensboro

IV. APPENDIXES

A Biographies of Speakers 143

B Participants List 161

C Bibliography 168

Preface

Recognizing that a capacity to innovate and commercialize new high-technology products is increasingly a part of the international competition for economic leadership, governments around the world are taking active steps to strengthen their national innovation systems. These steps underscore the belief that the rising costs and risks associated with new potentially high-payoff technologies, and the growing global dispersal of technical expertise, require national R&D programs to support new and existing high-technology firms within their borders. They also reflect the belief that shared facilities, coupled with geographical proximity, can facilitate the transition of ideas from universities and laboratories to private markets.

What is the impact of these initiatives for the competitive position of the United States? In a recent report, the National Academies warned that "this nation must prepare with great urgency to preserve its strategic and economic security," adding that "the United States must compete by optimizing its knowledge-based resources, particularly in science and technology, and by sustaining the most fertile environment for new and revitalized industries and the well-paying jobs they bring."[1]

Understanding the change in nature of these new institutions is a first step in understanding that the nature and terms of economic competition are shifting.[2]

---

[1]National Academy of Sciences/National Academy of Engineering/Institute of Medicine, *Rising Above the Gathering Strom: Energizing and Employing America for a Brighter Future,* Washington, DC: The National Academies Press, 2007.

[2]Kent Hughes has argued in this regard that the challenges of the 21st century require new strategies that take account of new technologies, new global competitors, as well as new national priorities concerning national security and the environment. See Kent Hughes, *Building the Next American*

U.S. policymakers need to be aware of the wide variety of innovation and competitiveness policies that many nations have adopted. These policies are designed to build research capacities and to acquire knowledge, and then to transition that knowledge directly to companies and support their development.

Some nations have developed well-financed and integrated national programs that are designed to enhance their economic growth, technical competency, and competitive position. Other national programs, while more modest in scale, provide essentially market-based incentives to encourage the transition of new technologies to the market. Yet, even these can have a significant impact on the terms of competition. While institutions and the scale of funding vary across the globe, a comparative perspective can help us understand what policies are succeeding and why, what we may learn from the experience of others, what existing U.S. programs might be enhanced, and what new initiatives might be launched. What is clear is that the terms of competition are shifting. Other nations are devoting very substantial resources to attract, develop, and nurture the industries of today and tomorrow. U.S. policy needs to be formulated with this understanding.

PROJECT OVERVIEW

Recognizing the importance of targeted government promotional policies relative to innovation, the Board on Science, Technology, and Economic Policy (STEP) is studying selected foreign innovation programs and comparing them with major U.S. programs. This analysis of Comparative Innovation Policy, carried out under the direction of an ad hoc Committee, includes a review of the goals, concept, structure, operation, funding levels, and evaluation of foreign programs designed to advance the innovation capacity of national economies and enhance their international competitiveness.

THE CONTEXT OF THIS REPORT

Since 1991 the STEP Board has undertaken a program of activities to improve policymakers' understanding of the interconnections among science, technology, and economic policy and their importance to the American economy and its international competitive position. The Board's interest in comparative innovation policies derives directly from its mandate.

This mandate has previously been reflected in STEP's widely cited volume, U.S. Industry in 2000, which assesses the determinants of competitive performance in a wide range of manufacturing and service industries, including

Century: The Past and Future of American Economic Competitiveness, Washington, DC: Woodrow Wilson Center Press, 2005, Chapter 14.

Box A
Innovation and Competitiveness

Innovation can be defined as the transformation of an idea into a marketable product or service, a new or improved manufacturing or distribution process, or even a new method of providing a social service. This transformation involves an adaptive network of institutions that encompass a variety of informal and formal rules, norms, and procedures—a national innovation ecosystem—that shape how individuals and corporate entities create knowledge and collaborate to bring new products and services to market.

If competitiveness is defined as the ability to gain market share by adding value better than others in the globalized economic environment, then the ability of entrepreneurs, venture capitalists, and scientists and engineers and others to collaborate successfully within a given innovation ecosystem gains significance. Recognizing this, policymakers around the world are supporting a variety of initiatives to improve cooperation within their national innovation ecosystems as a way of improving their national competitiveness. Science and technology parks are widely perceived as an effective mechanism to promote such partnerships

those relating to information technology.[3] The Board also undertook a major study, chaired by Gordon Moore of Intel, on how government-industry partnerships can support the growth and commercialization of productivity enhancing technologies.[4] Reflecting a growing recognition of the importance of the surge in productivity since 1995, the Board also launched a multifaceted assessment, exploring the sources of growth, measurement challenges, and the policy framework required to sustain the New Economy.[5]

The current study on Comparative Innovation Policy builds on STEP's experience to develop an international comparative analysis focused on U.S. and foreign innovation programs.

To open this analysis, the Committee held a symposium on April 15, 2005, which drew together leading academics, policy analysts, and senior policymakers

[3]National Research Council, *U.S. Industry in 2000: Studies in Competitive Performance,* David C. Mowery, ed., Washington, DC: National Academy Press, 1999.

[4]This summary of a multivolume study provides the Moore Committee's analysis of best practices among key U.S. public-private partnerships. See National Research Council, *Government-Industry Partnerships for the Development of New Technologies: Summary Report,* Charles W. Wessner, ed., Washington, DC: The National Academies Press, 2003. For a list of U.S. partnership programs, see Christopher Coburn and Dan Berglund, *Partnerships: A Compendium of State and Federal Cooperative Programs,* Columbus, OH: Battelle Press, 1995.

[5]National Research Council, *Enhancing Productivity Growth in the Information Age: Measuring and Sustaining the New Economy,* Dale W. Jorgenson and Charles W. Wessner, eds., Washington, DC: The National Academies Press, 2007.

from around the globe to describe their national innovation programs and policies, outline their objectives, and highlight their achievements.[6] Follow up symposia in Taipei and Tokyo in January 2006 focused on the evolution of the Taiwanese and Japanese innovation systems over the past decade. The Committee also convened a major conference in Washington in June 2006 that identified current trends in the Indian innovation system and highlighted the new U.S.-India innovation partnership.[7] This was soon followed by a symposium on "Synergies in Regional and National Innovation Policies in the Global Economy" held in Flanders, Belgium. This event reviewed European Union, national and regional innovation policies in Flanders, a region of Belgium, with a major university and research center with a strong commercialization record. Flanders is also home to IMEC, one of the leading microelectronics research facilities in the world and the flagship of Flemish technology policy.

A Joint Effort

This report captures the presentations and discussions of a March 2008 conference on best practices among science and technology research parks around the world. The conference was organized jointly with the Association of University Research Parks (AURP). By drawing on the AURP's expertise and contacts, the conference brought together leading figures from government, universities, and science and technology parks, both from the United States and around the world. The goal of the conference was to increase policymakers' understanding of the role of research parks as sources of innovation and regional growth, while also reviewing their contributions to government missions and to the commercialization of university research.

Parks Are a Diverse Phenomenon

An important characteristic of research parks is their diversity. Accordingly, the conference examined a broad range of research parks, including both university- and laboratory-based parks as well as the large-scale industrial models often undertaken in Asia and Europe. While recognizing the diversity of objectives and the differences in scope and scale of activity, the conference sought to identify common challenges faced by research parks both in the United States and abroad, including evaluation and the need for appropriate metrics.

Of course, no one-day conference can capture all aspects of this complex

[6]For a summary of this conference, see National Research Council, *Innovation Policies for the 21st Century*, Charles W. Wessner, ed., Washington, DC: The National Academies Press, 2007.

[7]For a summary of this conference, see National Research Council, *India's Changing Innovation System: Achievements, Challenges, and Opportunities for Cooperation*, Charles W. Wessner and Sujai J. Shivakumar, eds., Washington, DC: The National Academies Press, 2007.

phenomenon, and the conference did not focus, for obvious reasons, on failed parks. While there may be strategies and programs that do not work effectively, even within successful parks, this conference was focused on the practices of successful parks around the world.

ACKNOWLEDGMENTS

The Board on Science, Technology, and Economic Policy wishes to acknowledge the support of the Defense Advanced Research Projects Agency, the National Institute of Standards and Technology, the National Science Foundation, the Office of Naval Research, and Sandia National Laboratories. Both for the preparation of this conference and this volume, we are most appreciative of the support offered by the Association of University Research Parks and the University of Maryland.

This conference benefitted from the active collaboration and support of both academics and practitioners. We are most grateful to Professor Albert Link of the University of North Carolina at Greensboro for his inspiration and encouragement for the organization of this conference as well as his commitment to understanding the role S&T parks play in national innovation systems. We would also like to recognize the key contributions of Eileen Walker, Executive Director of the Association of University Research Parks, and Jackie Kerby Moore, the Executive Director of the Sandia Science and Technology Park, to the selection of outstanding speakers able to capture the diversity of parks around the world. Similarly, we would also like to recognize Professor Phillip Phan of Rensselaer Polytechnic Institute for his valuable suggestions of well-qualified speakers. Special thanks are also due to Robert Geolas of Clemson University and Brian Darmody of the University of Maryland for their contributions as well as to Michael Bowman, President of the Association of University Research Parks, for his support and leadership. This cooperative effort contributed a great deal to the scope and diversity of the representatives and to the quality of the conference deliberations.

With regard to the preparation of this report, we are indebted to Alan Anderson for his preparation of this meeting summary and to Sujai Shivakumar for his preparation of the draft introduction to this volume. Several members of the STEP staff also deserve recognition for their contributions to the preparation of this report, including Jeffrey McCullough and David Dierksheide for their role in preparing on an accelerated basis both the conference and report for publication.

NATIONAL RESEARCH COUNCIL REVIEW

This report has been reviewed in draft form by individuals chosen for their diverse perspectives and technical expertise, in accordance with procedures approved by the National Academies' Report Review Committee. The purpose of this independent review is to provide candid and critical comments that will assist the institution in making its published report as sound as possible and to ensure that the report meets institutional standards for quality and objectivity. The review comments and draft manuscript remain confidential to protect the integrity of the process.

We wish to thank the following individuals for their review of this report: Brian Darmody, University of Maryland; William Kittredge, U.S. Department of Commerce; Michael Luger, University of Manchester; Lora Lee Martin, California Council on Science and Technology; and Chachanat Thebtaranonth, National Science and Technology Development Agency, Thailand.

Although the reviewers listed above have provided many constructive comments and suggestions, they were not asked to endorse the content of the report, nor did they see the final draft before its release. Responsibility for the final content of this report rests entirely with the author and the institution.

Alan Wm. Wolff Charles W. Wessner

Summary*

Many nations are currently adopting a variety of directed strategies to launch and support the development of research parks, often with significant financial commitments and policy support. To meet the need for a better understanding of the scope and scale of programs overseas to support the growth and development of research parks and to improve our understanding of the scale and contributions of parks in the U.S. context, the National Academies convened an international conference on global best practices in research parks. This report of that conference captures the rich discussion of the diverse roles university- and laboratory-based research parks play in national innovation systems. The presentations of the conference participants demonstrate the range of objectives and the substantial differences in scope and scale of activity characterizing research parks around the world, while also identifying common challenges.

The conference included participants from countries as diverse as China, Singapore, India, the United Kingdom, France, Mexico, and Hungary. They described national efforts to develop research parks of significant scale and scientific and innovative potential. They noted that in many cases, these research parks are expected to generate benefits that go beyond regional development and job creation. As these participants observed, effective research parks have the potential to contribute to regional growth by facilitating innovation and forms

*Caveat: It is important to note that this summary does not represent the statements and conclusions of the NRC Committee on Comparative Innovation Policy, but simply summarizes key points raised by participants of the conference.

Box A
What is a Research Park?

Alternatively referred to as research parks, science parks, technology parks, technopoles, science centers, business innovation centers, and centers for advanced technology, there appears to be no singular characterization of a research park. The International Association of Science Parks defines a Science Park as "an organisation managed by specialised professionals, whose main aim is to increase the wealth of its community by promoting the culture of innovation and the competitiveness of its associated businesses and knowledge-based institutions. To enable these goals to be met, a Science Park stimulates and manages the flow of knowledge and technology amongst universities, R&D institutions, companies and markets; it facilitates the creation and growth of innovation-based companies through incubation and spin-off processes; and provides other value-added services together with high quality space and facilities."[a]

[a]Accessed at *<http://www.iasp.ws>* on January 22, 2009.

formation while enhancing the competitiveness of national firms, particularly in leading technological sectors.

Research parks are seen increasingly around the world as a means to create dynamic clusters that accelerate economic growth and international competitiveness. Specifically, research parks, of various sizes and types, are widely seen as an effective policy tool to realize larger and more visible returns on a nation's investments in research and development. Most research parks seek to encourage greater collaboration among universities, research laboratories, and large and small companies, in order to facilitate the conversion of new ideas into the innovative technologies for the market.[1] They are widely considered to be a proven tool to encourage the formation of innovative high-technology companies.[2] They are also seen as an effective means to generate employment and to make existing companies more competitive through cooperative R&D, shared facilities, and the benefits derived from co-location.[3]

S&T Parks are a rapidly growing phenomenon and an increasingly common tool of national and regional economic development. They are designed to:

[1]See, for example, the presentation by Dr. M. S. Ananth of the Indian Institute of Technology-Madras in the Proceedings section of this report.

[2]See, for example, the presentation by Zhu Shen on the Chinese strategy concerning research parks in the Proceedings section of this report.

[3]See, for example, the presentation by Yena Lim on Singapore's strategy in the Proceedings section of this report.

— Facilitate the cooperation that generates higher returns on existing investments in R&D and large-scale research facilities;[4]

— Meet the special needs of high-tech industries for infrastructure and associated services;[5] and

— Achieve critical mass in terms of co-located research facilities and staff.[6]

In the United States, innovative clusters and parks have developed as a result of government action and private initiatives, and in some cases around a government-funded laboratory. Research Triangle Park was built on national investments in major research universities in the Research Triangle area and has benefitted from patient state support and funding.[7] The Sandia Science and Technology Park, located near Sandia National Laboratories in New Mexico and the NASA Research Park, located at the NASA Ames facility in California are two examples of successful research parks that advance the missions of federal laboratories and research facilities while contributing to regional economic growth and mission-related innovation.[8] As a result of this success, innovation clusters, such as North Carolina's Research Triangle (and less directly Silicon Valley), are now being emulated around the world, often on a significantly larger scale.[9]

The United States has led the way in park creation and the generation of high-tech clusters. Yet on a global basis, the United States is not making comparable efforts, nor are federal programs supporting regional and state efforts to the same degree.[10] Investments by the world's leading nations in research parks reflect an appreciation of their capacity to spur knowledge-based growth and a national commitment to enhance technological competitiveness through innovation.[11] While research parks such as those at NASA Ames and Sandia have recorded significant progress, and new federal initiatives such as that of the National Cancer Institute are underway, the potential of research parks appears

[4]See, for example, the presentation by John Niederhuber of the National Cancer Institute in the Proceedings section of this report.

[5]See, for example, the presentation by Richard Stulen of the Sandia National Laboratories in the Proceedings section of this report.

[6]See, for example, the presentation of Ilona Vaas of the Hungarian National Office for Research and Technology in the Proceedings section of this report.

[7]See the presentation of Rick L. Weddle on the Research Triangle Park in the Proceedings section of this report.

[8]See the presentations of Richard Stulen of Sandia National Laboratories and the presentation of Pete Worden of NASA Ames Research Center in the Proceedings section of this report.

[9]See, for example, the remarks of Jane Davies of the Manchester Science Park, United Kingdom, in the Proceedings section of this report. She notes that U.S. research parks serve as models overseas.

[10]See the keynote remarks of Senator Jeff Bingaman in the Proceedings section of this report. Senator Bingaman has previously sought to introduce legislation to support state and local efforts to develop research parks.

[11]See, for example, the presentations on research park activities in China, India, Singapore, and France in the Proceedings section of this report.

to be less appreciated by policymakers and the public in the United States. In the United States, support for research parks is principally undertaken by state and local governments with limited support by the federal government.[12]

While research parks are highly varied, the conference nonetheless captured some common elements characteristic of successful research parks. These "best practices" include the presence of:

- **Champions:** Committed champions who match sustained, high-level attention with significant support for the growth and development of a research park.[13]
- **Leadership:** Effective leadership and professional management to facilitate networking among the entrepreneurs, researchers, investors, and others within and around the research park's innovation ecosystem.[14]
- **Funding:** Designated and sustained public funding and active private participation, combined with effective public policies to support companies that seek to convert ideas into innovations and innovations into products for the market.[15]
- **Bridging institutions:** Such as the North Carolina Board on Science and Technology, that preserve the vision of the research park over the long period it takes for parks to mature and become successful.[16]
- **Soft infrastructure:** This term captures the positive human capital built over many years of public investments in education and skills training, public policies that encourage an entrepreneurial culture, and the presence of networks among professionals.[17]
- **Metrics:** Effective metrics to help management set clear goals and, over time, gauge the effectiveness of the research park.[18]

---

[12]See the remarks by Prof. Phillip Phan in the Proceedings section of this report.

[13]Dr. Richard Stulen drew attention to the role that effective high-level champions like Senator Bingaman have played in the growth of New Mexico's Sandia Science and Technology Park.

[14]Clear goals, capable management, and sustained support are essential for the effective development of research parks, as documented in recent reviews by the National Research Council of the research parks affiliated with the NASA Ames Research Center and Sandia National Laboratories. See National Research Council, *A Review of the New Initiatives at the NASA Ames Research Center*, Charles W. Wessner, ed., Washington, DC: National Academy Press, 2001, and National Research Council, *A Review of the Sandia Science and Technology Park Initiative,* Charles W. Wessner, ed., Washington, DC: National Academy Press, 1999.

[15]In his keynote address, President Barker of Clemson University emphasized the instrumental role played by the State of South Carolina as well as private donors like BMW in providing sustained support and the scale of funding needed to provide the ICAR "instant scale and instant density." See the presentation of James Barker in the Proceedings section of this report.

[16]See the presentation by Robert McMahan in the Proceedings section of this report.

[17]Describing the case of Singapore, Ms. Yena Lim noted that the government has over 40 years built a strong education system, raising the value of learning and rewarding scientific and engineering excellence. See the presentation by Yena Lim in the Proceedings section of this report.

[18]See the presentations by Albert Link of the University of North Carolina and William Kittredge of the Department of Commerce in the Proceedings section of this report.

I

INTRODUCTION

Research Parks in the 21st Century

Science and technology research parks are seen increasingly as a means to create dynamic clusters that accelerate economic growth and international competitiveness. A concept that is now 60 years old, research parks are widely believed to encourage greater collaboration among universities, research laboratories, and large and small companies, providing a means to help convert new ideas into the innovative technologies for the market.[1] In this way, research parks are recognized to be a "proven tool to create successful new companies, sustain them, attract new ones—especially in the science, technology, and innovation sector—and make existing companies more successful through the use of R&D."[2]

Today, countries as diverse as China, Singapore, India, and France are among those undertaking substantial national efforts to develop research parks of significant scale and scientific and innovative potential. In many cases, these research parks are expected to generate benefits that go beyond regional development and job creation. Indeed, to the extent that research parks are effective, they have the potential to shift the terms of global competition, not least in leading technological sectors. For example, as Senator Bingaman noted in his keynote remarks at

[1]The first research park was established in Menlo Park, California, in 1948. Early successful parks, established in the 1950s and early 1960s, include the Stanford Industrial Park (est. 1953) in California, Research Triangle Park (est. 1958) in North Carolina, and Waltham Industrial Center (est. 1954) in Massachusetts. See Rachelle Levitt, ed., *The University/Real Estate Connection: Research Parks and Other Ventures*, Washington, DC: Urban Land Institute, 1987. See also Roger Miller and Marcel Cote, *Growing the Next Silicon Valley: A Guide for Successful Regional Planning*, Toronto: D. C. Heath and Company, 1987.

[2]See the presentation of Dr. Ilona Vass, director of the Hungarian National Office for Research and Technology, in the Proceedings section of this report.

Box A
Making Research Parks a Priority

"Many countries have been able to use the mechanism of S&T parks to greatly advance their technological capabilities. We have not given it nearly the emphasis of other countries. I would like to see the government provide more assistance to states to make research parks a priority."

Senator Jeff Bingaman, Keynote remarks at
the 2008 National Academies conference on S&T Research Parks

the National Academies conference that are summarized in this volume, research parks focused on software design in India have supported that nation's emergence as a global leader in software design and services.[3]

Yet, while investments by the world's leading nations in research parks reflects an appreciation of their capacity to spur knowledge-based growth and enhance technological competitiveness through innovation, this potential of research parks appears to be less well understood by policymakers and the public in the United States.

To better understand the role that research parks can play as sources of innovation, regional growth, and national competitiveness for the United States and to document recent developments in the growth of research parks around the world, the National Academies' Board on Science, Technology, and Economic Policy (STEP) partnered with the Association of University Research Parks (AURP) to bring together leading figures from governments, universities, and research parks from the United States and around the world.[4]

[3]In the early 1990s, the Indian government established a network of national software technology parks that provided broadband connectivity based on satellite and fiber technology, single-window clearance system to software exporters, and incubation services. These research parks have helped India generate a substantial return on national investments on research and training in science and engineering. Today, the software industry is dominated by globally competitive champions from India, including Tata Consultancy Services (TCS), Infosys, and Wipro Technologies, each of which has generated revenues in excess of US$1 billion a year in recent times. See Tilman Altenburg, Hubert Schmitz, and Andreas Stamm, "Breakthrough: China's and India's Transition from Production to Innovation," *World Development* 36(2):325-344, February 2008. Senator Bingaman's keynote address is summarized in the Proceedings section of this report.

[4]The conference, which was organized in cooperation with AURP, took place on March 13, 2008, at the National Academy of Sciences in Washington, DC, AURP is an association of planned and operating research parks around the world. Its mission is "to promote the development and operations of research parks that foster innovation, commercialization and economic competitiveness in a global economy through collaboration among universities, industry and government."

The conference was timely, reflecting the growth of research parks around the word in terms of numbers, scale of operations, and the participation of many leading global corporations. The conference was also timely with respect to recent legislation introduced by Senators Bingaman, Bunning, and Pryor to provide federal assistance to states that are seeking to develop research parks.[5]

This report of that conference captures the rich discussion of the diverse roles that university- and laboratory-based research parks play in national innovation systems. It also captures the central role of government support in the growth of research parks in countries as varied as China, Singapore, India, Hungary, France, the United Kingdom, Mexico, and the United States. The conference presentations demonstrated the range of objectives and the substantial differences in scope and scale of activity characterizing research parks around the world, while also identifying common challenges.

ADDRESSING THE 21ST CENTURY INNOVATION CHALLENGE

The global proliferation of research parks reflects the recognition by national and regional governments that future economic growth and competitiveness lies in developing a robust knowledge economy. The proliferation of research parks highlights the need for a better understanding by U.S. policymakers of the ambitious objectives and sheer scale of these programs in terms of their structure, operation, and funding levels. Indeed, these partnership programs and supporting policy measures are of great relevance to the United States and other countries both for their potential impact on national competitiveness and for the policy lessons they offer.

A New Urgency

From the U.S. perspective, one cannot assume that American preeminence as an innovative nation is assured in this new global competition. As the National Academies noted in its recent report, Rising Above the Gathering Storm, "this nation must prepare with great urgency to preserve its strategic and economic

[5]In 2004, through Senator Bingaman's introduction of S. 2737, "The Science Park Administration Act of 2004," and again in 2007 through Senator Pryor's introduction of S. 1373, "The Building a Stronger America Act," the U.S. Congress considered, but did not pass, a bill to provide grants and loans to states and local authorities for the development and construction of university parks. Implicit in these bills is the assumption that R-S-T parks are an important element in the U.S. national innovation system and as such should be fostered because of both the knowledge-based and employment-based spillovers that will result. The Science Park Administration Act of 2004 was reintroduced in S. 1581 as the Science Park Administration Act of 2005. Albert Link, in his overview of the academic literature (in this volume) on research S&T parks, notes that "in the United States, public investment at state universities is used to underwrite the formation and development of R-S-T parks," which is, by definition, a limited source of investment.

security. Because other nations have, and probably will continue to have the competitive advantage of low-wage structure, the United States must compete by optimizing its knowledge-based resources, particularly in science and technology, and by sustaining the most fertile environment for new and revitalized industries and the well-paying jobs they bring."[6]

If the United States is to respond to this challenge, its policymakers must understand what is happening at the cutting edge of policies to advance science, technology, and innovation around the world. In his remarks, James Turner, the Chief Counsel of the House Committee on Science at the time, affirmed that U.S. policies must address these new realities if the nation is to compete successfully in the 21st century.

To contribute to policymakers' understanding of global developments in innovation, the National Academies' STEP Board has undertaken a comparative review of policies and programs around the world designed to stimulate knowledge-based growth. The study highlights the growth around the world of investments in research and development as well as new public-private partnerships to foster the collaboration needed to translate ideas born in the laboratory into competitive new products for the marketplace.[7]

The Role of Public-Private Partnerships

Public-private partnerships are widely seen as a tool required to address today's innovation imperative. By stimulating cooperative research and development among industry, government, and universities, partnerships can play an instrumental role in introducing new technologies to the market.[8] A recent review by the National Academies of best practices among U.S. innovation partnerships shows that cooperative research and development among industry, universities,

---

[6]National Academy of Sciences/National Academy of Engineering/Institute of Medicine, *Rising Above the Gathering Storm: Energizing and Employing America for a Brighter Economic Future*, Washington, DC: The National Academies Press, 2007.

[7]For an overview of some of the key innovation programs currently underway around the world, see National Research Council, *Innovation Policies for the 21st Century*, Charles W. Wessner, ed., Washington, DC: The National Academies Press, 2007. Other volumes in the series include National Research Council, *India's Changing Innovation System: Achievements, Challenges, and Opportunities for Cooperation*, Charles W. Wessner and Sujai J. Shivakumar, eds., Washington, DC: The National Academies Press, 2007, and National Research Council, *Innovative Flanders: Innovation Policies for the 21st Century*, Charles W. Wessner, ed., Washington, DC: The National Academies Press, 2008.

[8]Not all policy observers accept the view that partnerships can play an important, even determinant, role in bringing technologies to market. Some would argue that most innovations come from the private sector acting alone and that government's role in supporting innovation is limited. They believe that the vitality of the U.S. economy rests almost exclusively on the dynamism of the private sector. These perspectives ignore the real-world gaps and irregularities in information that often preclude successful collaboration within the private sector. The Nobel Committee in awarding the 2001 Nobel Prize in Economics to George A. Akerlof, A. Michael Spence, and Joseph E. Stiglitz recognized their pathbreaking analysis of these information asymmetries.

Box B
The Role of Public-Private Partnerships

"Partnerships facilitate the transfer of scientific knowledge to real products; they represent one means to improve the output of the U.S. innovation system. Partnerships help by bringing innovations to the point where private actors can introduce them to the market. Accelerated progress in obtaining the benefits of new products, new processes, and new knowledge into the market has positive consequences for economic growth and human welfare."[a]

Government-Industry Partnerships for the Development of New Technologies
A Report of the National Academies

[a]National Research Council, *Government-Industry Partnerships for the Development of New Technologies*, Charles W. Wessner, ed., Washington, DC: The National Academies Press, pp. 2-3.

and government laboratories can work if properly designed, effectively led, and adequately funded.[9] The growing importance of collaboration in bringing research to the market and the positive role of federal support for innovative small companies is documented in a recent independent study of the changes in the United States innovation landscape over the past four decades.[10]

Research parks are a type of public-private partnership that "fosters knowledge flows—often between park firms and universities and among park firms—and

[9]A National Research Council Committee led by Gordon Moore concluded that "Public-private partnerships, involving cooperative research and development activities among industry, government laboratories, and universities, can play an instrumental role in accelerating the development of new technologies from idea to market." See National Research Council, *Government-Industry Partnerships for the Development of New Technologies: Summary Report*, Charles W. Wessner, ed., Washington, DC: The National Academies Press, 2003, p. 23.

[10]See Fred Block and Matthew Keller, "Where Do Innovations Come From? Transformations in the U.S. National Innovation System, 1970-2006," The Information Technology and Innovation Forum, July 2008. Accessed at <*http://www.itif.org/files/Where_do_innovations_come_from.pdf*>. The authors analyze a sample of innovations recognized by *R&D Magazine* as being among the top 100 innovations of the year over the past four decades. They find that while in the 1970s almost all winners came from corporations acting on their own, more recently over two-thirds of the winners have come from partnerships involving business and government, including federal labs and federally funded university research. Moreover, in 2006 77 of the 88 U.S. entities that produced award-winning innovations were beneficiaries of federal funding.

contributes to regional economic growth and development."[11] These partnerships enhance, both formally and informally, the efficiency of innovation within park firms, universities, and national laboratories.

As Professor Albert Link of the University of North Carolina at Greensboro and others have noted, scientific research is often characterized by high "spill-overs" of knowledge.[12] In such cases, private investors may not be able to capture the benefits of research at sufficient levels to justify investment in that research. Public investments in facilities such as research parks can reduce the costs faced by individual firms and thus increase the willingness of universities and private firms to perform research. The result of this research spills over to other firms in the park and in the local and national economy.[13] Many countries around the world as well as some states here in the United States have adopted measures to lower costs for firms by providing commercial facilities that enhance the research process, lower its cost, and where appropriate speed its dissemination.

Box C
Parks as a Nexus of Innovation

"Research Parks appear to be an excellent place to cross the Valley of Death between invention and the marketplace."

Dr. Lawrence Schuette, Office of Naval Research

UNDERSTANDING RESEARCH PARKS

Alternatively referred to as research parks, science parks, technology parks, technopoles, science centers, business innovation centers, and centers for advanced technology, there appears to be no singular characterization of a research park.[14] The International Association of Science Parks defines a Science Park as

[11]See Albert N. Link, "Research, Science, and Technology Parks: An Overview of the Academic Literature," in this volume. As Link notes these definitional characteristics are emphasized by President Mote of the University of Maryland and President Barker of Clemson University in their conference presentations.

[12]Albert N. Link and John T. Scott, "The Economics of University Research Parks," *Oxford Review of Economic Policy* 23(4):661-674, 2007.

[13]See the paper by Albert Link, "Research, Science, and Technology Parks: An Overview of the Academic Literature," in this volume.

[14]There appears to be no uniformly accepted definition of a Science Park in the academic literature although some analysts have attempted to distinguish between Innovation Centers, Science Parks and Research Parks. For a survey of the literature, see Hans Löfsten and Peter Lindelöf, "Science Parks and the Growth of New Technology-based Firms—Academic-industry Links, Innovation and Markets," *Research Policy* 31(6):859-876, August 2002.

"an organisation managed by specialised professionals, whose main aim is to increase the wealth of its community by promoting the culture of innovation and the competitiveness of its associated businesses and knowledge-based institutions. To enable these goals to be met, a Science Park stimulates and manages the flow of knowledge and technology amongst universities, R&D institutions, companies, and markets; it facilitates the creation and growth of innovation-based companies through incubation and spinoff processes; and provides other value-added services together with high quality space and facilities."[15]

While this definition is comprehensive, the fact is that research parks around the world vary considerably in terms of their mission and scale. Their physical location can range from an exclusively urban settings—such as Singapore's emerging Biopolis or the IIT-Madras Research Park in Chennai, India—to more spread out facilities—such as the Research Triangle Park in North Carolina—or even to very large sites like the Zhongguancun and Zhangjiang Science Parks in China. In practice, the terms "science park" and "technopole" are used most commonly in Europe, the term "technology park" is more prevalent in Asia, while the term "research park" is preferred in the United States and Canada.[16] We use the term "research park" in this paper.

Box D
Understanding the Diversity of Research Parks

"If you've seen one research park . . . you've seen one research park."

Prof. Albert Link, University of North Carolina at Greensboro

Defining research parks in the most general terms, Professors Michael Luger and Harvey Goldstein refer to them as "organizational entities that sell or lease spatially contiguous land and/or buildings to businesses or other organizations whose principal activities are basic or applied research or development of new products or processes."[17] In complement, Professor Paul Westhead emphasizes

[15]Accessed at <*http://www.iasp.ws*> on January 22, 2009.

[16]E. J. Malecki, *Technology and Economic Development,* New York: John Wiley, 1991.

[17]Michael I. Luger and Harvey A. Goldstein, *Technology in the Garden*, Chapel Hill, NC: University of North Carolina Press, 1991, p. 5. This volume was a seminal study of the research parks phenomenon. With support from the Economic Development Administration of the U.S. Department of Commerce, Professors Luger and Goldstein have recently updated their study. See M. Luger and H. A. Goldstein, *Research Parks Redux: The Changing Landscape of the Garden*, Washington, DC: U.S. Economic Development Administration, 2006. This update reflects the evolution of the perceived role of research parks over the past two decades. Reflecting the evolution of the role of research parks, the 1991 study focused on the distribution of park benefits among different groups in the population, while the 2006 study focuses on the capacity of research parks to stimulate innovation and regional development.

Box E
Research Parks and the Nonlinearity of Innovation

The term, "basic research" refers to the advancement of fundamental knowledge and the theoretical understanding of the relations among variables. A researcher's curiosity, interest, and intuition often drive its exploratory nature. By contrast, the term "applied research" describes the use of accumulated theories, knowledge, methods, and techniques, for a specific purpose. Thus separated, basic research logically precedes applied research, which in turn precedes the development of an idea into a practical application.

This linear model of innovation (see Figure E-1) creates the misimpression that increasing public and private investments in research will automatically result in greater commercialization, strengthening, in turn, a nation's competitiveness in global markets.[a] As Dr. Robert McMahan of the North Carolina Board of Science and Technology noted at the conference, academic R&D creates innovation capacity, but does not by itself raise economic output or create the fast-growing "gazelle" firms that spur new growth.

In practice, distinctions between basic and applied research are often obscure, with the different stages of research intermixing frequently. (Figure E-2 shows a more complex "feedback model" of innovation.) This is particularly true for research and development in biotechnology and electronics, where fundamental research can be undertaken alongside work intended to develop new products. This can also be the case in areas where public- and private-sector partners collaborate in order to develop greater insight into areas of interest.

In this regard, a key advantage of research parks is that they offer scientists,

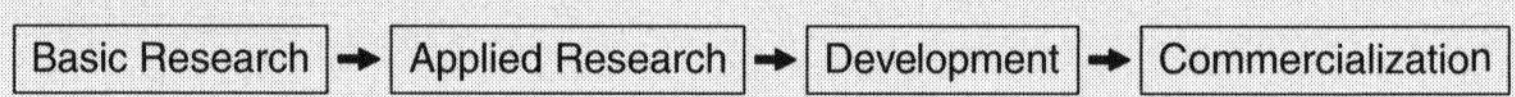

FIGURE E-1 A linear model of innovation.

the collaborative and catalytic environment that research parks provide that helps transform ideas born from research into products for the market.[18]

[18]P. Westhead, "R&D 'Inputs' and 'Outputs' of Technology-based Firms Located In and Off Science Parks," *R&D Management* 27(1):45-62, 1997. In addition to infrastructure investments in buildings and equipment and the employment of researchers and engineers, research parks often house incubation programs to provide resources that enhance the founding of new technology-based firms, although the effectiveness of these incubation programs vary. See K. F. Chan and T. Lau, "Assessing Technology Incubator Programs in the Science Park: The Good, the Bad and the Ugly," *Technovation* 25(10):1215-1228, October 2005. The authors find that the benefits required by technology founders at different stages of development are varied, and therefore, the general merits that are claimed by incubators as useful to technology start-ups are debatable.

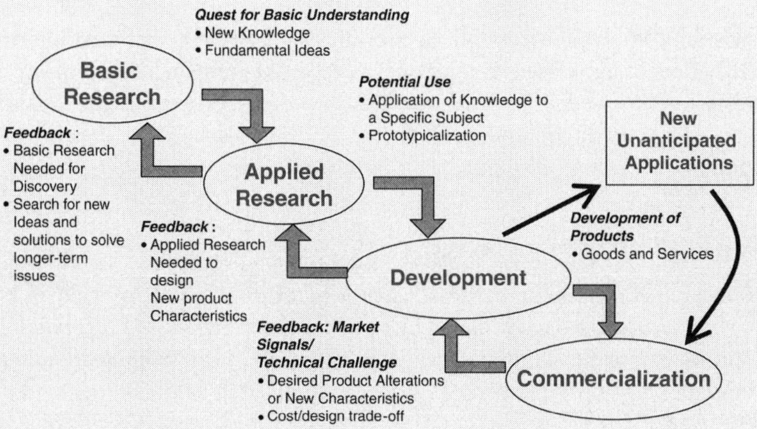

FIGURE E-2 A nonlinear feedback model of innovation.

entrepreneurs, venture capitalists, and others a supportive environment that encourages collaboration across disciplines and feedback across the different stages of innovation.[b]

---

[a]While elegant, it is easy to forget that this linear theoretical model overlooks the complex collaborations that characterize real-world innovation processes. For an analysis of such collaborations, see Dries Faems, Bart Van Looy, and Koenraad Debackere, "Interorganizational Collaboration and Innovation: Toward a Portfolio Approach," *Journal of Product Innovation Management* 22(3):238-250, 2005. Drawing from an empirical study of Belgian firms, the authors highlight relevance of "adopting a portfolio approach to inter-organizational collaboration within the context of innovation strategies."

[b]Andrew C. Inkpen and Wang Pien, "An Examination of Collaboration and Knowledge Transfer: China-Singapore Suzhou Industrial Park," *Journal of Management Studies* 43(4):779-811, 2006.

The conference examined two important types of research parks—University Research Parks and Laboratory Research Parks—with both types seen as versatile and adaptable to differing missions and environments.

University Research Parks

Professors Albert Link and John Scott define a university research park as a "cluster of technology-based organizations that locate on or near a university campus in order to benefit from the university's knowledge base and ongoing research. The university not only transfers knowledge but expects to develop knowledge more effectively given the association with the tenants in the research park."[19] Highlighting their role in fostering regional development, AURP refers to university research parks as property-based ventures that promote a university's research and development activities as well as local economic development.[20] AURP believes that effective parks can aid in the transfer of technology and business skills between university and industry teams, encourage the creation of startups, and promote technology-led economic development for the community or region.[21]

Adding to a University's Prestige

Speaking at the conference, Professor Albert Link of the University of North Carolina at Greensboro noted that while parks gain knowledge and prestige from their university partners, the universities gain as well. Association with a successful park can bring multiple benefits to a university such as:

- Higher publication rates;
- More successful patenting activities;
- A greater ability to hire eminent scientists; and
- An ability to garner larger extramural grants.

He noted that provosts of universities associated with parks, interviewed in the course of his research, indicated that one result of successful university-industry partnership in the context of a research park is a shift in the university curricula toward more applied research. This also helps graduate students in science and engineering find employment following their degrees. As a result, the university as a whole becomes more interested in innovation. Faculty who mentor students are more attuned to the needs of industry. The change in the structure of the curriculum is reflected by improved student placements.[22]

---

[19]Albert N. Link and John T. Scott, "U.S. University Research Parks," *Journal of Productivity Analysis* 25(1):43-55, 2006. For a thorough review of the Research Triangle Park, see Albert Link, *Generosity of Spirit: The Early History of the Research Triangle Park*, Research Triangle Park, NC: University of North Carolina Press for the Research Triangle Park Foundation, 1995.

[20]AURP Website, accessed at <*http://www.aurp.net/about/whatis.cfm*>, July 11, 2008.

[21]Research universities are increasingly recognized as an essential tool for regional economic development. Research universities are a key source of the fundamental scientific research that drives innovation, and this includes research in finance and economics and the dissemination of business methods. Most patent applications coming from industry research laboratories cite fundamental, exploratory research done at research universities. Research universities also produce the cadre of scientists and engineers who conduct research and development. They educate the workforce that brings innovation to fruition in commercial markets. See the speech by G. Wayne Clough, "The Role of the Research University in Fostering Innovation," The Americas Competitiveness Forum, June 12, 2007. Accessed at <*http://smartech.gatech.edu/bitstream/1853/22317/1/oop07-004_Americas_Compet_Forum_6-07.pdf*>.

[22]See also Albert N. Link and John T. Scott, "U.S. Science Parks: The Diffusion of an Innovation and Its Effects on the Academic Missions of Universities," *International Journal of Industrial Orga-*

Box F
Research Parks and the Missions of the 21st Century University

The wealth and competitiveness of nations increasingly depends on their ability to convert new knowledge into products for the market. Universities have traditionally been devoted to the production of new knowledge and its dissemination through their mission to educate students.

Universities have, in fact, always been active in both research and in its application. With the acceleration of knowledge-based competition by both established and emerging economies around the world, institutions of higher education increasingly find that they have no choice as to *whether* to be entrepreneurial or not. Indeed, the issue today for many of today's university leaders more squarely concerns *how* the university can assist in accelerating the commercialization of new knowledge and the development of local economies.[a]

One consequence of the conference was to highlight the growing role of research parks in helping universities balance their 21st century missions in education, research, and commercialization.

[a]Robert E. Litan and Lesa Mitchell, "Should Universities be Agents of Economic Development?" *Astra Briefs,* Vol. 7, Nos. 7-8, Summer 2008.

Expanding the University's Reach

Dr. C. D. Mote, President of the University of Maryland, noted that a university research park is an "essential tool for institutions with an entrepreneurial and innovative culture that hoped to benefit from complicated partnerships on a global scale." One need is the physical space and facilities afforded by a park that are not available on a university campus. Another is the need to be able to do proprietary or classified research, which is not easily done in an academic environment.[23] A third is the ability to accommodate a large off-campus work force to achieve the clustering, resonance, and mutual energy of people working and thinking together. Finally, a research park can uniquely bring in the many nonuniversity

nization 21:1323-1356, November 2003.

[23]This is a particularly important consideration for the University of Maryland, which is located near several federal agencies including the National Oceanic and Atmospheric Administration (NOAA), the National Institute for Standards and Technology (NIST), the Department of Homeland Security, the National Security Administration, the Food and Drug Administration, the National Institutes of Health, NASA, the Small Business Administration, and the National Archives. Some of these agencies conduct classified research, and many research universities have policies prohibiting classified research on campus, but not at affiliated research parks.

interests required by complex partnerships, including private industry services, security for confidential business activities, and government facilities.

Such considerations, he noted, underlay the creation of M Square, the research park of the University of Maryland. M Square, he said, seeks to attract people who would benefit from being close to the university and who would bring benefits to the university as well. Asked about public policies needed to launch research parks like M Square successfully, Dr. Mote replied that timely investment in the startup phase is essential. Even though M Square had benefitted from research agreements with the National Oceanic and Atmospheric Administration and the federal government's Intelligence Advanced Research Projects Activity, he noted that prompt financial support from the state and federal government would have reduced startup difficulties and accelerated M Square's progress.

Promoting Research Collaboration

In his conference keynote, President James Barker of Clemson University emphasized the role that research parks can play in promoting collaboration. There will always be "eureka moments" in the labs where scientists work, he said, but we are now discovering that collaboration can be a tremendous competitive advantage. He said that Clemson has worked hard to make collaboration a part of campus culture, including the establishment of the Clemson University—International Center for Automotive Research (CU-ICAR) on a 250-acre campus. Four years ago CU-ICAR was 250 acres of undeveloped land along the I-85 corridor between Charlotte and Atlanta, with no master plan, no business plan, no curriculum, and no funding, he noted. Today CU-ICAR includes a 90,000-square-foot graduate engineering center that has attracted world-class faculty who hold well-funded endowed chairs.

Aligning Incentives and Missions of Universities and Research Parks

The missions of the 21st century university—to commercialize as well as to educate and expand knowledge—are not always well aligned. For example, as Dr. Phillip Phan noted in his remarks at the conference, university scientists have not traditionally viewed the commercialization of research as their priority, although this attitude has changed substantially in the past 20 years.[24] Incentives facing the tenants of a University Park and the research and commercialization missions of the university can also be misaligned. However, as in other collective human endeavors, these university-industry interactions can be improved through

[24]The propensity of U.S. university faculty to work directly with industry on research activities that lead to patents is changing as incentive structures facing faculty change. See Stuart D. Allen, Albert N. Link, and Dan T. Rosenbaum, "Entrepreneurship and Human Capital: Evidence of Patenting Activity from the Academic Sector," *Entrepreneurship Theory and Practice* 31(6):937-951, 2007.

Box G
Aligning Incentives in a University Research Park: The IIT-Madras Credit System

Addressing the issue of how to align the actions of the tenants of the research park with the objectives of the university, Dr. M. S. Ananth, Director of the Indian Institute of Technology-Madras (IIT-M), emphasized the importance of incentives that promote university-industry interaction. "Traditionally," he said, "we have found that one of the problems in situating parks near campuses is that they focus only on the real estate and they don't interact with the university. We are working hard to change that."

To address this challenge, Dr. Ananth has instituted a system of credits that tenants have to earn to remain in the IIT-M research park. Each company must earn a minimum number of credits by interacting with the university. The credit system, he noted, is designed to promote entrepreneurial activity, intersector interaction, and partnerships. Companies are given credits for participating in research and development projects with IIT-M, serving as consultants to IIT faculty, earning royalties, sponsoring doctoral and masters students, serving as adjunct faculty in order to teach and mentor graduate and undergraduate students, and providing part-time employment to students.

developing new rules of collaboration. Illustrating this point at the conference, Dr. M. S. Ananth, Director of the Indian Institute of Technology-Madras discussed how he has sought to align the incentives of research park tenants with IIT's mission. (See Box G.)

Laboratory Research Parks

Like universities, national laboratories are also repositories of knowledge and scientific aptitude and thus represent promising nuclei for the growth of innovation clusters. The United States government has made and continues to make substantial investments in the nation's laboratories, which have developed a significant store of technology and talent; these laboratories have much to offer to the private sector.

Moreover, the laboratories themselves recognize that they cannot fulfill their mission in isolation, especially given today's rapid pace of innovation. To remain effective, laboratories such as Sandia, NASA Ames, and the National Cancer Institute understand that they must stay abreast of the rapid technological change taking place in the commercial arena. This means building and maintaining ties to the private sector. As leaders of these three laboratories noted in their conference presentations, one means of encouraging this mutually beneficial exchange has been through the development of research parks.

Stimulating Joint Research at Sandia Park

New Mexico's Sandia Science and Technology Park was founded in 1998 to attract industry in support of the mission of the Sandia National Laboratories. Drawing on the proximity to the exceptional expertise and infrastructure at Sandia, the goal of the park was to stimulate joint research and development opportunities, commercialize technologies, bring in new business, strengthen supplier-based "collaboratories," and foster regional economic development.[25]

According to Dr. Richard Stulen, Chief Technology Officer of Sandia Laboratories, the park is now home to 27 companies, 2,113 employees, 18 buildings, 897,000 square feet of occupied space, and 67 developed acres. Funds-in and in-kind services flowing from tenants to Sandia, such as Cooperative Research and Development Agreements (CRADAs) and licensing agreements, have totaled \$17.6 million, and Department of Energy/Sandia in-kind services to tenants (CRADAs) have totaled \$2.7 million.[26] In the other direction, contracts from Sandia procurement to tenants amount to \$244.5 million. "The funding goes both ways," said Dr. Stulen. "This demonstrates what partnerships can do."

Box H
The Importance of Champions

"Parks don't just happen. They require energy, devotion, passion from leaders—not only of the institution but also of the region."

Dr. Richard Stulen, Chief Technology Officer of Sandia National Laboratories

Speaking at the conference, Dr. Stulen emphasized that expertise and infrastructure are not sufficient to ensure the success of a research park. Successful laboratory-based parks also require one or more high-level champions, people who care and have the ability to direct resources continuously to the park. He praised Senators Domenici and Bingaman of New Mexico for that leadership as well as other political leaders at the state and local levels. Success for Sandia Park, he noted, is rooted in a variety of active public-private partnerships involving agencies at the federal, state, and county levels, as well as the leadership of the Sandia Park.

[25]National Research Council, *Industry-Laboratory Partnerships: A Review of the Sandia Science and Technology Park Initiative*, Charles W. Wessner, ed., Washington, DC: National Academy Press, 1999.

[26]A CRADA, or Cooperative Research and Development Agreement, is a written agreement between a federal research organization and one or more federal or nonfederal parties (collaborators) to work together as partners on a research project of mutual interest.

Drawing in Talent at NASA Research Park

Since 1998, NASA has sought to develop the NASA Research Park (NRP) with the goal of creating a world-class, shared-use science and technology campus for government, academia, nonprofits, and industry. Located in the heart of California's Silicon Valley, NRP helps NASA achieve its mission by providing economical access to technological capabilities external to NASA. Some of the nation's leading technology companies are neighbors to NRP, including Google, Hewlett-Packard, Apple, and Intel. The park's goal is to draw in tacit knowledge from the exceptional technological and entrepreneurial community around Ames, while serving both as a source of trained personnel and as a conduit for laboratory innovations.[27]

According to the NASA Ames Center Director, Dr. Pete Worden, the park's mission is also to generate revenue for the Ames Center, using an enhanced use leasing program to convert underutilized lands. Finally, the park is seen as a tool to raise political and public support for NASA by providing benefits to the local economy, public education, and the U.S. scientific base. "We want to boost NASA's prominence before Congress and the public, strengthen community ties and investments, and broaden NASA's relationships with industry."

Given the progress in achieving these goals, Dr. Worden sees the NRP as a successful initiative, one that continues to progress. The scale is significant. As Dr. Worden observes, "We already have more than 40 industry and 14 university partners onsite. We have several million square feet of old Navy facilities, including the dirigible hangers. We have many successful R&D collaborations and spinoffs, support from our congressional delegation and the U.S. government, a NASA-approved business plan, and an approved environmental plan." The initial concept of the NASA Ames Park was reviewed by the National Academies' STEP Board in 1999.[28] The park has made great progress since then, exceeding expectations and enacting NASA's plans with remarkable effectiveness.

Accelerating Cancer Research at NCI-Frederick

In his conference presentation, Director John Niederhuber of the National Cancer Institute (NCI) noted that the word "cancer" really refers to hundreds of diseases—including, for example, many types of breast cancers and many types of colon cancers—all genetically different. Understanding this complexity will

[27]The National Research Council's Board on Science, Technology, and Economic Policy, under the aegis of the Gordon Moore Committee, worked closely with the NASA leadership in providing a comprehensive review of the plans for the NRP and the challenges to be addressed. See National Research Council, *A Review of the New Initiatives at the NASA Ames Research Center*, Charles W. Wessner, ed., Washington, DC: National Academy Press, 2001.

[28]National Research Council, *Industry-Laboratory Partnerships: A Review of the Sandia Science and Technology Park Initiative*, op. cit.

require more knowledge about genetic changes and differences than about the organ site. To cope with the growing research intensity of modern cancer studies, he said, NCI is developing a research park close to its NCI-Frederick campus in Maryland to bring together researchers and practitioners, thus creating a wider knowledge base and greater opportunities for industry partners to develop marketable applications of research.[29]

Dr. Niederhuber listed the specific advantages of a research park for NCI. First, he noted that some 82 percent of the NCI budget is spent extramurally to support research at research universities. The Institute has realized that it could make this spending even more effective by creating a science and technology research park close to the Frederick campus. Second, the proximity of the campus would allow NCI to improve productivity and reduce the costs of drug development and diagnostics. Third, a park might enable execution of larger and more complex experiments and provide a better environment for training the next generation of people to work in cancer-related technology development. Finally, he noted, a research park can offer incubator and think-tank space to integrate the work of multiple companies and research teams with significant R&D programs in engineering, software, and imaging.

To bring this plan into reality, NCI is negotiating with developers to create a park of about 230,000 square feet. According to the plan, NCI will be the anchor of the park, but it will support the broader mission of technology development for biological research.

RESEARCH PARKS AND ECONOMIC GROWTH

In addition to their role in advancing the research and commercialization missions of universities and national laboratories, research parks are widely seen as catalysts for the development of innovative clusters that support rapid economic growth.[30]

Developing Clusters of Innovation

The observation that firms tend to group together to profit from shared expertise and services and the development of mutual trust has encouraged interest in fostering industry clusters to enhance regional development. Examining industrial clusters from the perspective of business strategy, Professor Michael Porter has pointed out that "the enduring competitive advantages in a global economy lie

---

[29]Established in 1971 by President Nixon, NCI-Frederick was established as a rapid response site to develop new technologies to support the "War on Cancer." It was given FFRDC status in 1975, which allows NCI to establish contractual relationships in streamlined fashion. Of the 38 FFRDCs, NCI-Fredrick is the only one devoted to biological research.

[30]E. J. Malecki, *Technology and Economic Development*, op. cit.

increasingly in local things—knowledge, relationships, motivation—that distant rivals cannot match."[31] Moreover, as Professor AnnaLee Saxenian has explained, greater geographic proximity encourages repeated interaction that helps build the mutual trust needed to sustain cooperation and speed the continual recombination of knowledge and skill. The importance of trust emerging through repeated face-to-face interactions has led Professor Saxenian to observe that "paradoxically, regions offer an important source of competitive advantage even as production and markets become increasingly global."[32]

In the United States, such clusters have often developed around a government-funded nucleus; one example is the high-technology industries that emerged and grew around the government laboratories and major universities in the Boston area. In other cases (e.g., Silicon Valley) multiple private industries interacting with a major university, and irrigated with substantial and sustained federal funding, created powerful developmental synergies.[33] In contrast to the relatively spontaneous emergence of these innovation clusters, a third approach to the development of innovation clusters is through the deliberate creation of research parks.[34] The co-location of creative activity within the concentrated geographical area of a research park can help create a "community of innovation" needed to transfer new ideas from universities and national laboratories to the marketplace.[35] Today, successfully created innovation clusters, such as North Carolina's Research Triangle, are being emulated around the world, often on a larger scale.

Box I
Five Factors Behind Successful Research Parks

- A strong science and industry base.
- The availability of finance.
- The presence of entrepreneurs.
- The presence of trust networks at an individual level.
- The opportunity for collaboration among universities, businesses, and other organizations.

[31]Michael E. Porter, "Clusters and the New Economics of Competition," *Harvard Business Review* 76(6):77-90, 1998.

[32]See AnnaLee Saxenian, *Regional Advantage: Culture and Competition in Silicon Valley and Route 128*, Cambridge, MA: Harvard University Press, 1994, p. 161.

[33]See Martin Kenney, ed., *Understanding Silicon Valley: The Anatomy of an Entrepreneurial Region,* Stanford, CA: Stanford University Press, 2000.

[34]See Yun-Shan Su and Ling-Chun Hung, "Spontaneous vs. Policy-driven: The Origin and Evolution of the Biotechnology Cluster," *Technological Forecast and Social Change,* 2008.

[35]Elayne Coakes and Peter Smith, "Developing Communities of Innovation by Identifying Innovation Champions," *The Learning Organization: An International Journal* 14(1):74-85, 2007.

Reviving Regional Economic Growth

Reviving growth in an economically disadvantaged region is a frequent focus of policymakers. Research parks are often adopted as one tool to restore a region's vitality. One example is the Manchester Science Park (MSP) in the United Kingdom, which was created explicitly with the goal of stimulating the regional economy and creating jobs. Manchester was badly weakened by the downturn in manufacturing of the 1980s, suffering job losses and economic distress. In her conference presentation, Ms. Jane Davies of MSP noted that representatives of Manchester government, the university, and the commercial sector, drew specifically on the model of Research Triangle Park in proposing a new science park for their city, raising an initial capital investment of 210,000 Pounds.

Indeed, North Carolina's Research Triangle Park is seen widely as a successful example of how a research park can reverse the fortunes of a region faced with economic decline and a shrinking job base due to decreasing manufacturing concentrations.[36] Describing the long-term economic impact of the Research Triangle Park on the regional economy, Mr. Rick Weddle, president of the Research Triangle Foundation, noted that the per capita income growth in Raleigh-Cary and Durham were far below the state average and national averages before the park was formed. In contrast, he noted, the per capita income of the region today significantly exceeds the U.S. average and far exceeds the North Carolina average. "In the 1960s it was one of the poorest regions in the southeastern United States, and today is among the wealthiest regions in the southeast."

Employment growth began at Research Triangle Park in the mid-1960s, a time when less than 12 percent of the employment in the region was in high-tech industries. After climbing slowly to around 10,000 jobs in the 1970s, the employee population grew rapidly during the 1980s and has continued to grow since then. Mr. Weddle projected continued growth to about 45,000 science and technology jobs in the park by 2016, spread among some 160 high-technology firms. This Research Triangle Park example of regional initiative, backed by a sustained commitment of resources and interest, buttressed by effective leadership, and—above all—coordinated through effective public-private partnering, is a model for the efforts of many countries to generate the dynamic new firms that will provide the growth and jobs of the future.

SUPPORT FOR PARKS AROUND THE WORLD

Science parks, in their many different forms, now exist in most parts of the world and are seen as a proven policy tool to spur economic growth and enhance technological competitiveness. To this end, as Senator Bingaman noted in his

---

[36]For a comprehensive history of RTP, see Albert N. Link, *A Generosity of Spirit: The Early History of the Research Triangle Park*, op. cit.

keynote address, they benefit from significant financial and policy support from national and state governments.

The United States remains an exception in this regard, where support for research parks is principally undertaken by state and local governments with only limited participation by the federal government. While many state governments are experimenting with technology zones to support research parks and technology incubators, and to increase technology-led economic development clusters, some advocates believe that the U.S. government should pursue a more comprehensive strategy to capture the potential benefits of research parks for economic growth and national competitiveness.[37]

Reflecting this growing interest, the conference reviewed the scale and mechanisms of support for a variety of research parks from around the world.

China's Support for Parks

China is frequently seen as one of the foremost practitioners of the research parks strategy for economic and regional development. China's large science and technology industrial parks symbolize that nation's strong determination to grow and become internationally competitive through significant national and regional investments in science-based economic development.[38] A central theme of China's recent Five Year Plans has economic development driven by technological progress, and large-scale research parks are a widely used mechanism to carry out this goal.[39] (See Figure 1.) Both the absolute number and scale of Chinese research parks are remarkable. China's 54 state-level science and technology industrial parks are designed to help develop the industrial base for rising industries in electronics and information technology, new materials, and biomedicine.

Aggressive intervention by national and local governments to create and grow large-scale research parks are a hallmark of Chinese policy. The state's role

---

[37]See Association of University Research Parks, "The Power of Place: A National Strategy for Building America's Communities of Innovation," Tucson, AZ: Association of University Research Parks, 2008. Access at *<http://www.aurp.net/meet/The_Power_of_Place.pdf>*.

[38]For a review of China's science and technology industrial parks, see Susan M. Walcott, *Chinese Science and Technology Industrial Parks*, Aldershot, UK: Ashgate Publishing, 2003. Walcott distinguishes three types of research parks in China. Perhaps the most well known are the *multinational development zones* such Shenzhen, Dongguan, and Suzhou, which emphasize the role of transnational corporations as growth engines. Another type is the *multinational learning zones* of which Shanghai is considered a leading example. Lastly, *local innovation learning zones* rely more on domestically generated technology with some interactions with foreign companies. Examples include Xian, which draws heavily on local university resources and China's defense production industry.

[39]Kazuyuki Motohashi and Xiao Yun, "China's Innovation System Reform and Growing Industry and Science Linkages," *Research Policy* 36:1251-1260, 2007.

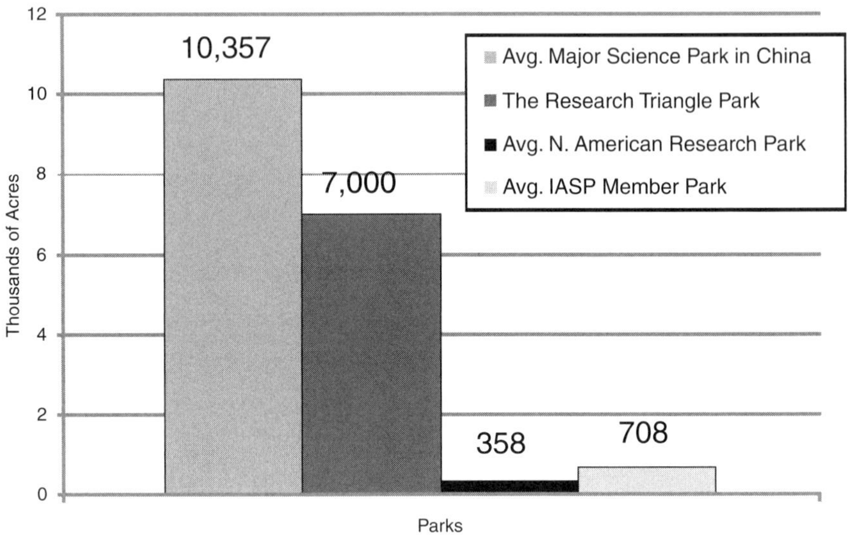

FIGURE 1 Research parks in comparative perspective—an issue of scale.
SOURCE: Presentation by Richard Weddle in the Proceedings section of this volume. "Average North American Research Park" data are from "Characteristics and Trends in North American Research Parks: 21st Century Directions," commissioned by AURP and prepared by Battelle, October 2007; "Average IASP Member Park" data are from the International Association of Science Parks annual survey, published in the 2005-2006 International Association of Science Parks directory.
NOTE: The scale of China's investments in research parks may be comparable to the massive efforts undertaken in the United States during the Cold War in building national laboratories. To the extent that these more commercially oriented investments are successful, they may well have a significant impact on the competitive position of Chinese industry.

in the growth of a park like the Zhangjiang High-Tech (ZHT) Park is illustrative.[40] Beginning almost from a clean slate, the Chinese authorities encouraged more than 30 research institutions to team up with research and development centers of multinationals to anchor the park site. These large research centers were joined by about 200 small and medium-sized Chinese biotechnology companies.

Outside ZHT, Shanghai Jiao Tong University and Fudan University contribute to the park's 8,600 strong workforce of scientists and researchers. But ZHT also benefits from national policies that attract Chinese overseas scientists back home. Policies planned for these returnees include low rent, tax breaks, and

[40]See Yun-Shan Su and Ling-Chun Hung, "Spontaneous vs. Policy-driven: The Origin and Evolution of the Biotechnology Cluster," op. cit.

assistance with living needs. At the end of 2004, there were 253 such returnees in to the ZHT park alone, and this trend is accelerating across China.[41]

The Chinese government is also a major financial supporter for biotechnology companies in ZHT. In addition to grants from the National Technology Innovation Fund for small and medium-sized businesses, the government established the Shanghai Pudong New Area Venture Fund to attract additional venture capital. In 2006, this amounted to more than $2.5 billion in venture funding for the ZHT park.

Box J provides brief descriptions of three of China's large-scale research parks, including ZHT.

While China provides a remarkable example of state support for research parks as a tool to promote national targets in technological progress, other nations are also providing significant support for their research parks. At the conference, representatives from several countries, including Singapore, France, and Mexico described major efforts to create and support research parks as a part of their nation's growth and development strategies.

Singapore's Parks Strategy: Investing for the Future

An island state with limited natural resources, Singapore has regularly reinvented and transformed itself though strategic and farsighted investments. Moving from labor-intensive production in the 1960s and 1970s, to skill-intensive production in the 1980s, to technology-intensive manufacturing in the 1990s. Singapore's GDP per capita has grown rapidly from US$512 in 1965 to US$35,640 in 2006.

Addressing the challenge of sustaining this growth, Dr. Yena Lim of the Singapore Agency for Science, Technology & Research (A*STAR) noted at the conference that her country is committed to investing in science and technology for the long-term to "ensure that the Singapore economy has the capability to reinvent itself and keep its international competitiveness." Towards this end, the Singapore government has allocated a five-year budget of S$13.55 billion (nearly US$10 billion) to strengthen its research and development base, especially in the area of biotechnology.

Research parks are central Singapore's growth strategy. Dr. Lim described Singapore's Biopolis project, planned as "the biomedical hub of Asia," a city within a city intended for scientists, researchers, and entrepreneurs. Located in central Singapore, it is designed to attract scientists from all over the world who will come for the quality of scientific research and the cosmopolitan work environment. Another park, Fusionopolis, located about half a mile from the Biopolis, is designed to be an integrated and comprehensive work-live-and-play

[41]China Daily, "China luring 'sea turtles' home," December 18, 2008. The recent U.S. financial crisis appears to be accelerating the trend of repatriating Chinese professionals and scholars.

Box J
Mega Parks in China

Noting that China has more than 54 state-level economic and technological development zones, and 53 national high-tech development zones, Dr. Zhu Shen of BioForesight described the development of three large science parks as a leading examples of Chinese policy and achievement.

Zhongguancun Science Park in Beijing

The park hosts over 20,000 enterprises and 950,000 employees, receiving total income of 850 billion Yuan (about US$124 billion). More than 800 enterprises have income exceeding 100 million Yuan. Of the industries represented in the park, the majority (56.6 percent) are classified as information technology, 12.5 percent as "new energy," 12.3 percent as biomedicine, 9.4 percent as advanced manufacturing, and 8.4 percent as new materials. The park has attracted almost 10,000 "sea turtles" (Chinese scientists who return home after acquiring skills abroad) that have set up 4,200 companies in Zhongguancun Science Park.

Shanghai Zhangjiang Hi-Tech Park

Established in July 1992 in the middle of Pudong New Area, the park comprises the Technical Innovation Zone, Hi-Tech Industry Zone, Scientific Research and Education Zone, and Residential Zone. The Zhangjiang High-Tech Park emphasizes three major areas of innovation: life science, which accounts for about 50 percent of revenues; software; and information technology. Its corporate tenants in the life sciences include six of the world's top ten pharmaceuticals and information technology companies. Chinese tenants include more than 60 small-molecule drug development companies, 35 medical device and diagnostics firms, and more than 15 traditional Chinese medicine companies. The park now accounts for 25 percent of Shanghai's GDP, 50 percent of foreign trade, and 30 percent of foreign investment. Of 25 square kilometers, 17 are already developed, hosting more than 3,600 companies, more than 140 of them foreign, and more than 100,000 employees. Reflecting this impressive investment, Dr. Shen observed, "I don't think we can find a park like this in the U.S., at least not yet."

Suzhou Industrial Park

Established in 1994, Suzhou Industrial Park is a unique joint development between the Chinese and Singapore governments. Located 80 kilometers west of Shanghai, Suzhou has taken its place at the high-tech frontier of the global economy. In land area only 0.1 percent and in population 0.5 percent of China, it accounts for 2.3 percent of GDP, 1.5 percent of financial revenue, 10 percent of imports and exports, and 8.3 percent of foreign investment. Of the Fortune 500 companies, 113 have set up operations in Suzhou.

environment. Opening in October 2008, Fusionopolis will have public- and private-sector labs, homes, service apartments, hotels, a shopping mall featuring smart-shopping technologies, food and beverage outlets, and an experimental theatre for art performances.

Both the Biopolis and the Fusionopolis, another research and development complex, are located within One-North, which is a larger area situated within Singapore's science and education talent belt, that also encompasses the National University of Singapore, the National University Hospital, part of the Nanyang Technological University, Singapore Science Park, and the Ministry of Education. One-North is well located—just 10 minutes from the city center and 20 minutes from the Changi International Airport. It is planned, Dr. Lim concluded, as a vibrant dynamic environment because "the research community must not be isolated." The ambitious premise of the design is to create "an ecosystem designed to nurture new ideas and push them quickly to reality."

Support for France's Grenoble-Minatec

France has an active policy to capture the benefits of its strong investments in research and to reinforce the now vibrant Grenoble high-technology cluster. Once a relatively small alpine community of 450,000, Grenoble was selected in 1957 as the site as 1 of 10 national centers dedicated to French nuclear research. Around the year 2000, as the potential for industry-government high-tech partnerships became clear, and as traditional nuclear engineering stagnated, the idea for Minatec was born. It offered a place and a rationale for formalizing new public-private partnerships within a major S&T park that could expand into new fields of cutting-edge research.

A research center focused on micro-nano technologies, Minatec is an extension of France's national laboratory system that has been redesigned to stimulate economic development.[42] Combining a research campus with a network of companies, researchers, and engineering schools, the success of the project in focusing on new products and miniaturized solutions for industry is such that the Grenoble region is now called France's Silicon Valley.

Speaking at the conference, Dr. David Holden of Minatec noted that the development of a research park in Grenoble has benefitted from a substantial 3.2-billion-Euro investment from the French government, with the local government adding about 150 million Euro, most of it to pay for infrastructure, such as highways and access roads. This investment, he said, has been more than paid

[42]Minatec members include CEA-Leti, one of Europe's largest microelectronics research institutes; INP-Grenoble, a French technical university; and industrial partners like the Crolles 2 Alliance of STMicroelectronics, Philips, and Freescale Semiconductor. Minatec facilities provide the space, tools, and expertise to companies locating their research and development projects in Grenoble. See Junko Yoshida, "Grenoble Lure: Un-French R&D," *EE Times,* June 12, 2006.

back in the form of corporate taxes over the four-year period, and the local government is still benefiting from a net positive of 1,000 technical jobs and perhaps three times as many support jobs.

Dr. Holden also noted that the Minatec model has influenced how other science and technology clusters in France are financed. France is trying to decentralize its financial system, with the regions promoting more of their own economic development. Competitive clusters, called "pôles de croissance," are creating groups to screen project applications and then ask for funding from the central government, a process put in place by President Sarkozy's government several years ago when he was Minister of the Interior. According to Dr. Holden, President Sarkozy is using the process at Minatec as the basis for the current French parks model. Under this system, Minatec received in just two years about 1.2 billion Euros (US$1.86 billion) for 113 projects and 315 million Euro (US$487 million) in financing from the government.

Developing a Research Park in Mexico: High-Technology Growth in Monterrey

Dr. Jaime Parada of Mexico's Research and Innovation Technology Park reported that a new research park is unfolding in Monterrey. The 175 acre research park draws on the dynamic industrial base of the city of Monterrey, which produces 11 percent of Mexico's manufacturing goods, equivalent to US$12.1 billion. Dr. Parada added that the province of Nuevo Leon also has multitiered higher educational institutions of good quality, including 93 colleges and universities that teach the skills and conduct research needed to support park activities. Monterrey Tech and the University of Monterrey, he said, are two of the most prestigious universities in Latin America, and the University of Nuevo Leon is ranked as one of the best state universities in Mexico. He noted that the province also has a strong base for research and development, with more than 1,500 researchers in public and private research institutions. Projected employment over the next five years is expected to grow to 3,500 researchers and engineers.

The research at the Monterrey research park is to be sustained with substantial support from the federal government of Mexico. According to Dr. Parada, this includes investment in infrastructure of $100 million, and investment in buildings and equipment of $150 million. Two business incubators have been designed, one for nanotechnology and one for biotechnology, at a cost of $20 million. The state's first seed and venture capital fund is being assembled by private partners, the government, and the national bank to a level of $30 million. In addition, he noted that Mexico absorbs 30 percent of annual R&D expenses as tax incentives for those who invest in research and development. Dr. Parada also reported that the Monterrey research park will launch in 2008 a new legal framework supporting research, development, and innovation, including a 25-year commitment of financial support.

CREATING SUCCESSFUL PARKS

Given the growth of new research parks around the world, how can we enhance the success of these ventures?[43] Several conference participants drew attention to a set of factors necessary for success.[44] Perhaps one of the most important factors is the presence and involvement of a large research university or laboratory supporting a critical mass of knowledge workers.[45] Also, key is availability of funding over a sustained period. Strong and committed leadership is also essential to facilitate and guide the development of the park's physical infrastructure and quality-of-life amenities. Finally, and not least, a successful park needs skilled entrepreneurs and managers. Talented and motivated individuals and teams in the private sector are needed to commercialize the knowledge generated.[46] If the benefits of a successful park are to be realized over the long term, a critical combination of these factors must be present, although they are not sufficient to ensure success.

Drawing on the experience of the Research Triangle Park, widely seen as a successful research park, Dr. Robert McMahan, the Science Advisor to the Governor of North Carolina, noted the importance of a policy environment that is patient, adaptable, and focused on commercialization.

The Importance of Patience

Dr. McMahan noted that when a small group of planners, over 50 years ago, began to create structures to leverage innovation resources within the state, North Carolina was "desperately poor," ranking 49th in the country in per capita income. Most of its population was involved in a few low-wage industries: tobacco, furniture, and textiles. What is distinctive about North Carolina's approach is that it recognized the importance of "patient structures," he said. For the first 10 years,

---

[43]Albert Link (in this volume) notes that research parks should not *a priori* be considered a primary element of a nation's innovation system. While successful research parks stimulate two-way knowledge flows between universities and industry, the conditions where such beneficial interactions can take place require further study.

[44]See, for example, the presentations by Ilona Vaas of Hungary, Jaime Parada of Mexico, and Yena Lim of Singapore in the Proceedings section of this report. See also David B. Audretsch, "The Prospects for a Technology Park at Ames: A New Economy Model for Industry-Government Partnership?" in National Research Council, *A Review of the New Initiatives at the NASA Ames Research Center*, op. cit., p. 119.

[45]Link and Scott note that "More research-oriented universities, as measured by RD100, have within their research parks a greater proportion of tenants that are university spin-off companies." See Albert N. Link and John T. Scott, "Opening the Ivory Tower's Door: An Analysis of the Determinants of the Formation of U.S. University Spin-off Companies," *Research Policy* 34:1106-1112, 2005.

[46]As Luger and Goldstein note, "Research parks will be most successful in helping to stimulate economic development in regions that already are richly endowed with resources that attract highly educated scientists and engineers." See Michael I. Luger and Harvey A. Goldstein, *Technology in the Garden*, op. cit., p. 184.

the park made little progress. But bridging institutions designed to institutionalize change, such as the North Carolina Board of Science and Technology, provided sustained support through successive elected state governments.

Experimentation and Adaptability

Dr. McMahan also noted that the North Carolina Board of Science and Technology experimented with organizational innovations, many of which have proved useful and have since served as models for newer research parks. Given the advantage of a long-term mandate, the Board has also been able to look ahead far enough to identify opportunities in new fields gaining momentum. He noted, for example, that the Board had urged the state to invest in biotechnology back when biotechnology was not yet a familiar term.

Focus on Commercialization

Given that a university-based park is part of a continuum of mixed infrastructure and other measures, Dr. McMahan noted that "entrepreneurship becomes critical to the success of the parks and the larger economy." Here, complementary public-private partnerships, like the Small Business Innovation Research (SBIR) program, can help maximize the state's investments in research parks while amplifying the impact of the federal government's investment in research.

SBIR is a federal program that offers competition-based awards to small high-technology firms with technically sound but commercially unproven ideas.[47] Because new ideas are by definition unproven, the knowledge that an entrepreneur has about his or her innovation and its commercial potential may not be fully appreciated by prospective investors.[48] This means that new ideas with commercial potential often do not attract sufficient private investment. SBIR awards provide this seed capital and, moreover, act as a signal to private venture capital markets, helping entrepreneurs secure the funds needed to bring new ideas to market.

Dr. McMahan noted that North Carolina recently instituted one of the country's most substantial support programs for SBIR award winners as part of entrepreneurship strategy. The state decided to award up to $100,000 in matching

[47]Created in 1982 through the Small Business Innovation Development Act, the Small Business Innovation Research (SBIR) is the nation's largest innovation program. For a comprehensive review of the concept and performance of this 25-year-old program, see National Research Council, *An Assessment of the SBIR Program*, Charles W. Wessner, ed., Washington, DC: The National Academies Press, 2008.

[48]Joshua Lerner, "Public Venture Capital," in National Research Council, *The Small Business Innovation Program: Challenges and Opportunities,* Charles W. Wessner, ed., Washington, DC: National Academy Press, 1999. For a seminal paper on information asymmetry, see Michael Spence, *Market Signaling: Informational Transfer in Hiring and Related Processes*, Cambridge, MA: Harvard University Press, 1974.

funds to each company that won an SBIR grant from the federal government. He noted that this approach has already succeeded in targeting high-potential small firms "and should be viewed as a powerful complement of research parks and other structures."

EVALUATING RESEARCH PARKS

As countries, regions, and cities around the world invest—sometimes on a massive scale—to create new research parks as a means to accelerate technological and economic transition, the task of measuring the contribution of research parks becomes more important. However, the empirical literature related to the formation and performance of universities and firms parks has been characterized as "embryonic."[49] The conference participants discussed the need for evaluating research parks, some of the challenges of evaluation, and the need to refine the metrics of evaluation.

Why Evaluate?

In his conference remarks, Professor Link noted that research parks should be evaluated to make them more accountable to the public that invests in them and to understand, improve, and measure the benefits they provide to universities, laboratories, and businesses and to contributing scientists and engineers.

Professor Link reported that more than 80 percent of research parks in the United States rely on government and/or university funds to develop park land and infrastructure. Given the size of this public expenditure, he noted that it is reasonable to expect that both "investors" and the local community will want accurate evaluations of parks.

He added that this strong need for accountability is reasonable because park directors are stewards of public money, and the public sector has a long history of being accountable for its use of public resources. He went on to note that accountability is an especially important issue for universities, which "are not known to be good managers of research parks." In this context, he suggested that evaluation exercises can actually help universities become better managers of research parks by helping them identify the multiple facets of the university that are affected by the park. Thus, another rationale for the evaluation of research parks is that they can reveal ways of increasing the efficiency of the science and commercialization efforts at the university and at independent companies.

---

[49]For a review of the empirical literature on research parks, see Albert N. Link, "Research, Science, and Technology Parks: An Overview of the Academic Literature," in this volume.

The Challenge of Evaluation

While noting that the need for evaluation is clear, Dr. William Kittredge of the Department of Commerce affirmed that developing effective metrics to measure the performance of research parks, and of economic development in general, is at best a "work in progress."

One fundamental limitation arises from the diversity of park characteristics and missions. As noted earlier, there is no commonly used definition of a research park. They can vary dramatically and—given their different goals, facilities, funding, and management structures—comparing one research park with another is a complex task. As even the small sample of parks referenced in this conference demonstrates, their location can vary considerably from an exclusively urban site, to a much more park-like setting. The immediate economic, political, and social environments surrounding each research park also vary. The actual organization, management, legal status, and size also vary significantly among research parks.

Thus, as Professors Luger and Goldstein have noted in their seminal analysis of research parks, "One of the conceptual difficulties is that there is no consensus about the definition of success. . . . The most commonly cited goals relate to economic development. But both the literature and our data from interviews with park developers, elected officials, university administrators, business leaders, and others confirm the existence of other goals, including technology transfer, land development, and enhancement of the research opportunities and capacities of affiliated universities."[50]

Another limitation arises from the absence of a systematic framework to understand the dynamic interactions among the various stakeholders and participants of a research park and how the nature of these interactions affects behavior and hence outcomes. As Professors Phillip Phan, Donald Siegel, and Mike Wright have observed in their recent review of the literature, there is a failure to understand the dynamic nature of research parks as well as that of the companies located on them.[51] Acknowledging this point, Dr. Kittredge noted in his remarks that qualitative measures may have an important role in park evaluation, adding dimensionality and nuance to more generalized, but limited, quantitative measurements.

Some Possible Metrics

These challenges notwithstanding, there are several metrics that can help gauge the relative success of a research park, as Professors Luger and Goldstein point out.[52] These include:

---

[50]Michael I. Luger and Harvey A. Goldstein, *Technology in the Garden*, op. cit., p. 34.

[51]Phillip H. Phan, Donald S. Siegel, and Mike Wright, "Science Parks and Incubators: Observations, Synthesis and Future Research," *Journal of Business Venturing* 20(2):165-182, March 2005.

[52]Luger and Goldstein have used a multiple case-study approach along with a quasi-experimental design where multiple areas with parks were compared to similar areas without research parks in order

- **Meeting the goals of legislation.** One plausible way to measure the success of research parks is to assess their performance against stated goals, as written into legislation and found in documents and interviews.
- **Return on public investments.** Direct expenditures by government on land acquisition and infrastructure development, financial inducements, and the opportunity cost of the land for research parks versus other types of uses can be compared against changes in the tax rolls and other measures of economic growth.
- **Enhanced firm performance.** This can be measured in terms of the change in income and corporate taxes collected by local, state, and federal governments as the result of the growth of successful businesses inside and outside the park, as well as in terms of net gains in jobs.[53]
- **Enhanced university performance.** As noted above, spillovers to the economy usually take the form of the creation of codified knowledge, which can be measured in terms of patents and publications. Spillovers can also be examined in other ways. Tenants often form research joint ventures with other firms in the park, and this can be tracked. Tenant companies may also provide benefits to the host university by sponsoring laboratories and professorships, hiring students, or associating themselves with co-patenting activity.
- **Value of the park to tenants.** Another kind of measure is the value of the park to tenant companies that benefit from the richness of the flow of knowledge between them and universities. For example, firms may seek the cachet of working in a successful park, which can benefit the host university, tenant firms, and the local community.

RESEARCH PARKS: LESSONS AND IMPLICATIONS

Research parks are widely seen as a proven partnership tool to increase the return on a nation's investment in research and development.[54] Found in both developed and developing nations, they are now a world-wide phenomenon. National governments are making very substantial investments in research parks to facilitate the commercialization of new technologies, to attract leading high-technology companies from around the world, to benefit from and contribute to university research and "market ready" students, and to create centers of regional and national economic development.[55]

---

to determine whether research parks did in fact meet their objectives. Michael I. Luger and Harvey A. Goldstein, *Technology in the Garden*, op. cit.

[53]Luger and Goldstein point out that one must also consider jobs created outside the park area because of the park's creation. Ibid.

[54]This point was emphasized by a number of speakers. See, for example, the remarks of Ilona Vaas of the Hungarian National Office for Research and Technology in the Proceedings section of this volume.

[55]See, for example, the presentation by Yena Lim on Singapore's initiatives and by Zhu Shen on China's significant investments in research parks in the Proceedings section of this volume.

By fostering a more robust interface between universities and laboratories and entrepreneurs and small and large businesses, research parks are seen as an effective policy tool to realize large and highly visible returns on a nation's R&D investments. Towards this end, many nations are adopting a variety of directed strategies to launch and support the development of research parks.[56] Box K summarizes the recent evolutions in the scope and form of these parks.

Box K
Research Parks in the 21st Century—Some Recent Trends[a]

- Research parks are no longer a developed world phenomenon. Parks can be found in more than 60 countries at all stages of development.
- Most research parks outside the United States are planned as part of a national strategy for industrial competitiveness.
- Many parks employ cluster-based recruitment and marketing methods, including tax incentives, training programs, and other industry-targeted services.
- Technological development at many research parks is increasingly integrated with university research, with faculty working with private firms and firms renting laboratories and incubator space in universities.
- Beyond research universities, community colleges and regional technical schools are increasingly participating in research parks.

[a]Based on M. I. Luger and H. A. Goldstein, *Research Parks Redux: The Changing Landscape of the Garden,* Washington, DC: U.S. Economic Development Administration, 2006.

Best Practices in Research Parks

What makes research parks effective? Several of the conference participants offered their own insights on some of the elements essential for success.[57] Some of their main points are summarized below.

Committed Champions

Successful parks generally begin with committed champions with long-term visions. In his presentation, Dr. Richard Stulen drew attention to the role that

[56]The presentations in this report provide a sample of efforts underway around the world to develop research parks. This includes specific parks in China, India, Singapore, Hungary, France, the United Kingdom, Mexico, and the United States.

[57]These views do not necessarily reflect those of the NRC Committee authoring this report.

effective high-level champions like Senator Bingaman have played in the growth of New Mexico's Sandia Science and Technology Park.

Park Leadership and Staff

Well-managed and well-staffed parks are essential to the development of research parks, as documented in recent reviews by the National Research Council of the research parks affiliated with the NASA-Ames Research Center and Sandia National Laboratories.[58] Professional managers can facilitate networking among researchers, entrepreneurs, venture capitalists, and other key players in the park's innovation ecosystem. Well-managed incubators located within research parks can also nurture startup companies. Dr. Mary Good of the University of Arkansas emphasized the importance of a research park's management and staff in her summary of the conference by underscoring that quality staffing is an essential element in success.

Committed Funding

Consequently, another ingredient essential to the emergence of research parks is the presence of committed funding. In his keynote address, President Barker of Clemson University emphasized the instrumental role played by the State of South Carolina as well as private donors like BMW in providing the funding needed to provide the ICAR "instant scale and instant density." "I can't overstate the importance of effective public policy and designated public funding," he concluded. "Otherwise, we would have had a great idea but no means of turning this idea into reality."

Bridging Institutions

As Dr. McMahon noted in his presentation, dedication to a vision, pursued over several decades by North Carolina policymakers, university administrators, and business leaders, has been instrumental in the successful development of Research Triangle Park. Given that the tenure of these leaders, especially elected officials, is short relative to the life of a research park, bridging institutions, such as the North Carolina Board of Science and Technology, play an important role in sustaining support for research parks over the long run.

---

[58]See National Research Council, *A Review of the New Initiatives at the NASA Ames Research Center*, op. cit., and National Research Council, *Industry-Laboratory Partnerships: A Review of the Sandia Science and Technology Park Initiative*, op. cit.

Relative Scale

Ms. Jane Davis of the Manchester Science Park noted that the impact of a research park is related to its scale relative to the local economy, available resources, and public expectations. Even relatively modest research parks, if adequately resourced and ably led, can have a major impact on a small community.

Soft Infrastructure

An innovative cluster is more than the sum of its capital investments. Describing the case of Singapore, Dr. Yena Lim noted that the government has over the long course built a strong education system, raising the value of learning and rewarding scientific and engineering excellence. Human capital, in terms of education and skills training, a vibrant entrepreneurial culture, and the presence of networks among professionals all contribute to an indigenous capacity to innovate. A well-designed and supported research park can then capitalize on these investments.

Implications for the United States

In her concluding remarks, Dr. Mary Good reflected on the acceleration in the pace of planning and development of research parks around the world. Time and patience are important, she observed, "but we have heard an undercurrent of urgency from essentially all of the participants." She noted further that sovereign states have decided to front-load economic development and to lean on their state-supported universities to contribute to national economic development of new technology-based companies.

Research parks are clearly a key element of economic development today, and we need to learn from others and adopt and adapt those lessons to the United States, just as other countries are adapting what they see as positive lessons from the U.S. experience.

> "If we can look at what we've learned here today and disseminate it quickly, we can do some good."
>
> Dr. Mary Good, University of Arkansas

Research parks in the United States can and must flourish beyond today's few dominant centers of science and technology expertise. This is a message, affirmed Dr. Good, which needs to be heard by governors and state legislatures around the nation. We need to capitalize on and support innovation around more of our universities throughout the country, she said, for the United States to continue to be as visible or viable in the next 50 years as we have been in the past 50 years.

II

PROCEEDINGS

Welcome

Charles Wessner
National Research Council

Dr. Wessner welcomed the participants and thanked the speakers, many of whom had travelled long distances to participate in this event. He noted that the National Academies' Board on Science, Technology, and Economic Policy (STEP) and the Association of University Research Parks (AURP) were jointly convening this symposium on global best practice in science and technology parks.

This symposium, he noted, is a key element in the ongoing efforts of STEP to identify and compare as appropriate best practices in innovation policies from around the world.[1] STEP is also interested in the synergies between state and federal programs, he said, as well as the synergies between the activities of foundations and regional economic growth in the United States.[2] The involvement of foundations is typically an American feature of development, he said, and one that must be examined to understand the innovation process in the United States. He noted that STEP has consistently sought to bring objective analysis to these exercises.

[1]See National Research Council, *Innovation Policies for the 21st Century*, Charles W. Wessner, ed., Washington, DC: The National Academies Press, 2007.

[2]STEP has initiated a study of "Best Practice in State and Regional Innovation Initiatives" in order to identify best practices with regard to their goals, structures, instruments, modes of operation, synergies across private and public programs, funding mechanisms and levels, and evaluation efforts.

THE IMPORTANCE OF EVALUATION

Effective innovation policies require an active program of evaluation and learning. But this is not always the case. "You'd be surprised at how often people move uncomfortably in their seats when you ask them, 'How do you know if you've succeeded?'" he said. "'And how could you measure that, or replicate it?' Too often we hear, 'Don't ask if it works or not—it's for a good cause.' We think that's the wrong approach."

A PORTFOLIO FOR INNOVATION

He called research parks an increasingly important element in a robust innovation ecosystem, but just one element. Another is the Small Business Innovation Research (SBIR) program, which STEP had also examined in depth. "Frankly," he said, "we were surprised at how effective that program is. A distinguished committee found it to be 'sound in concept and effective in practice,' and that has encouraged its renewal by Congress."[3]

THE INNOVATION IMPERATIVE

Dr. Wessner noted that as policymakers around the world recognize the importance of innovation for economic growth and national competitiveness, they are increasingly adapting public-private partnerships like SBIR and S&T Parks to their own national circumstances. Then, summarizing what he called the "innovation imperative, he drew out the following three points:

- Innovation is the key to maintaining a country's competitive position in the global economy.
- The importance of small businesses and universities in the innovation process is seldom recognized.
- Science and technology research parks have quickly become one of the most important catalysts of innovation.

Among the key issues to be addressed in the symposium, he said, is the evaluation of research parks and their role in commercializing government-funded research. Underscoring the fact that research parks are diverse, and all have different histories, goals, and structures, he cited, what he called, the "Link dictum," of Professor Albert Link of the University of North Carolina at Greensboro: "If you've seen one park, you've seen one park." At the same time,

[3]See National Research Council, *An Assessment of the Small Business Innovation Program,* Charles W. Wessner, ed., Washington, DC: The National Academies Press, 2008.

he said, this meeting offered a chance to identify some of the challenges common to all parks, including the need for more accurate assessment techniques.

Finally, he expressed his gratitude for the support and participation of the Office of Naval Research, the National Institute of Standards and Technology (NIST), the National Science Foundation, the University of Maryland, and Clemson University, the informal steering group that help plan the event, and the staff of the Academies and AURP for their professionalism and efficiency in organizing the event.

Keynote Address I

Introduction

Mary Good
University of Arkansas at Little Rock;
and Board on Science, Technology, and Economic Policy

Dr. Good of the National Academies STEP Board, welcomed the audience and expressed her honor and delight in welcoming Senator Jeff Bingaman to deliver a keynote address. She said she had worked with the senator for many years, describing him as a staunch friend of science and technology. "For many of the issues we talk about today, where we've had some good luck at the federal level, if you pull back the rug a little, you'll almost always find Jeff Bingaman involved somewhere in that endeavor."

Senator Bingaman, she said, truly understands that sustained federal investment in science and technology is essential to ensure the future of American innovation, and to maintain a competitive economy with high-wage job production. In particular, he had played a major role in the passage of the American Competitiveness Act, which recognized investments in innovation as a national priority, and urged continuing work to move the Act from the authorization stage to the appropriation stage.

She further praised his understanding that science-driven economic development is a major component of national security, in addition to the more explicit role of science in underpinning defense technologies. She noted that Senator Bingaman, as chair of the Energy and Natural Resources Committee, had written numerous bills to move the nation toward energy security by spurring innovation in renewable and environmentally friendly technologies. Most relevant to the topic at hand, she said, is his sponsorship of the Science Parks Administration Act, which is designed to strengthen the nation's science park infrastructure and establish a science park venture capital program.

Jeff Bingaman
United States Senate

Senator Bingaman (D-New Mexico) described his state's significant participation in research parks. New Mexico has five science parks which, as of March 2007, employed more than 2,600 people in well-paying, high-technology jobs representing about 105 organizations. The capital investment in these parks amounted to nearly half a billion dollars—a figure, he said, that is expected to grow.

He said that his attention was first drawn to science and technology parks when he visited several well-known parks that had been established in Taiwan, Hong Kong, and India. He was struck by the commitment of the regions to their parks and the economic success generated by businesses in the parks. For example, he attributed Taiwan's high standing in the global marketplace largely to the success of its Hsinchu Science Park, which was established in 1980. As of 2004, the park held more than 100,000 technically trained people, two major universities, 385 companies, and six national laboratories. It generated gross revenues of more than US$32 billion, according to government figures. Taiwan has more recently built parks in the central and southern parts of the country as well, and gross revenue from all three parks is projected to exceed $66 billion per year.

He said that in Hong Kong, the government has invested nearly half a billion dollars to construct ten buildings in its park, holding more than one million square feet of office and laboratory space. Among the topics emphasized in the park are integrated circuit design, photonics, biotechnology, and information technology. In 2005, he visited Hyderabad and Bangalore in India and witnessed the success of the research parks there, especially those focused on software design. India was then supporting more than 1,000 companies in 44 software parks. At Infosys, perhaps the most successful software development company in India, he asked the CEO to explain the hiring process. During the previous year, the company received 1.2 million applications for jobs. They gave all applicants a standardized test, reducing the number to 300,000, then interviewed 30,000 of them and hired only 10,000. "That," commented the senator "is a select group of people."

SOME COMMON FEATURES OF PARKS

Science parks now exist in most parts of the world, but with many different structures. Nonetheless, said Senator Bingaman, they do share some important common features. First is a government commitment to provide a first-class infrastructure capable of accommodating different levels of science-based companies. Second, parks try to bring together companies of similar interests so they can mutually reinforce one another along the supply chain. Third, many encourage a system of "one-stop shopping" for companies that need basic services and inventions, providing assistance in streamlining government approvals and systems for loans and loan guarantees. Fourth, parks usually provide tax incentives, generally waiving them completely during the early years when capital expenditures are high.

NEW LEGISLATION ON S&T PARKS

In many countries, the success of parks depends heavily on the participation and commitment of government at every level. To increase this commitment at the federal level in the United States, the senator introduced the Science Park Administration Act in 2004 to create mechanisms for more effective science parks. He reintroduced the bill in 2005, joined by Senator Bunning of Kentucky, and in the present Congress, Senator Pryor of Arkansas has taken the lead, with Senator Bingaman as co-sponsor.

"Many countries have been able to use the mechanism of S&T parks to greatly advance their technological capabilities," concluded Senator Bingaman. "We have not given it nearly the emphasis of other countries. I would like to see the federal government provide more assistance to states that want to make research parks a priority."

Keynote Address II

Introduction

Mary Good
University of Arkansas at Little Rock;
and Board on Science, Technology, and Economic Policy

Dr. Good next invited Dr. Dan Mote, president of the University of Maryland (UM), to deliver the second Keynote Address. She described him as another "friend of science and technology" and said that he has strengthened S&T programs throughout his institution during his tenure. His goal, she said, was to "lead the state in development of a high-tech economy, especially in ICT, bioscience, biotechnology, and nanotechnology—in other words, the emerging technologies that will drive the 21st century." She noted also his accomplishments in expanding university partnerships with corporate and federal laboratories, and his role in bringing to College Park the first research park sponsored by the People's Republic of China outside the mainland. He has also helped create M Square, a new research park affiliated with the university.

C. D. Mote, Jr.
University of Maryland

In the spectrum of research parks, Dr. Mote began, M Square is a "small endeavor" very much focused on the University of Maryland. Nonetheless, he said, it served a function at the university that could not be served in any other way.

He said that the University of Maryland placed a high value on innovation and entrepreneurship in everything it does, and that this culture leads directly to a partnership mode of operation. Having healthy partnerships expand assets, he said, and the domains in which an institution can work. Partnerships also increase the speed with which people can execute their operations.

Today's partnerships between universities, industries, and government are increasingly globalized and exist in many new combinations. In a university, a partnership might form at the faculty or college level, or as a consortium of universities. In industry, partnerships may form at the level of multinationals, local firms, or industry associations. In government, partnerships may involve cities, states, regions, and entire countries, raising the complexity of the average partnership far beyond what it was 10 or 20 years ago. The research park, he said, fills the need for a multipurpose structure where partners from multiple sectors can interact in the physical proximity so important to innovation. For a university, a park can bring about a powerful expansion of the university's mission.

AN UNUSUAL PARTNERSHIP WITH CHINA

The openness of the University of Maryland has led to one unusual international partnership based on the research park concept—the UM-China Research Park. This partnership originated in 2002, when Chu Wang Wa, then Minister of Science and Technology, was looking for a site in the United States to develop a university-industry-government partnership. The University of Maryland was selected, agreeing to provide services from both the School of Engineering and the School of Business, along with a university outreach group. Staff members were sent from China to help develop the park, and last summer the university sent two vice presidents and two deans on a tour to recruit Chinese companies. Working in Beijing, Shanghai, and Suzhou under the sponsorship of the Ministry of S&T, Maryland recruited about ten companies for the new park.

This partnership has produced a number of additional programs, said Dr. Mote, including a flexible training program for Chinese middle-management executives. These executives may design their studies from a wide range of topics, including the U.S. government, management, Olympic games management, democracy, banking, and finance. About 1,000 people have come for two months to a year of training.

The university also hosts a Confucius Institute Program, created after China's admission to the World Trade Organization. Set up by the Chinese national Office for Teaching Chinese as a Foreign Language, its purpose is to help other countries learn about Chinese culture and language. After a pilot program at Maryland in 2004, the program is now held at more than 30 sites in the United States and 200 around the world, all of them under the auspices of the UM-China partnership. It offers degree programs in Chinese in business, engineering, journalism, criminal justice, agriculture, and public policy.

PARTNERSHIP WITH THE STATE AND INDUSTRY

The entrepreneurial and innovative spirit of the university has led to activities with many partners. One of these is the Maryland Technology Enterprise Institute, a partnership between UM and the state of Maryland. For more than 20 years, the Institute has carried out economic development activities for the state. One is the state's most successful business incubator, he said, which was created in 1985. The Institute also runs a bioprocess scale-up facility, which allows companies to scale up their processes from bench-level bioproduction to production scale. Another major activity is the Maryland Industrial Partnerships, a research program that brings faculty members together with companies with the objective of creating greater value for products. In addition, a venture accelerator helps move technologies developed by faculty and students to commercialization, and a Technology Startup Boot Camp brings hundreds of would-be entrepreneurs to weekend camps to learn principles of entrepreneurship.

These and other related activities have brought about $17 billion in revenue to the state of Maryland since 1984, some $14 billion of this from research activities. The Institute is credited with creating about 7,000 jobs, a significant figure for a small state.

As might be expected, the business school at the university is integrally involved in such programs. The Dingman Center for Entrepreneurship, founded in 1986, is an entrepreneurial institute that focuses on enterprise creation. Major features include a networked program to improve capital access, an executive mentoring program, and the weekly "pitch Dingman" competitions in which competitors pitch their business idea for the chance of winning a $500 prize.

A major asset of UM, said Dr. Mote, is its location near many federal agencies and institutes. Many of these have research and training partnerships at or near the M Square Park, including the National Oceanic and Atmospheric Administration (NOAA), NIST, Department of Homeland Security, National Security Administration, Food and Drug Administration, National Institutes of Health, NASA, Small Business Administration, and National Archives.

THE NEED FOR A RESEARCH PARK

With such assets and other entrepreneurial activities, he asked, why did the university need a research park? Dr. Mote said that a park is an "essential tool for institutions with an entrepreneurial and innovative culture that hoped to benefit from complicated partnerships on a global scale." One need is the physical space and facilities afforded by a park that are not available on a university campus. Another is the need to be able to do proprietary or classified research, which is not easily done in an academic environment. A third is the ability to accommodate a large off-campus work force to achieve the clustering, resonance, and mutual energy of people working and thinking together. Finally, a research park can uniquely bring in the many nonuniversity interests required by complex partnerships, including private industry services, security for confidential business activities, and government facilities.

Such considerations underlay the creation of M Square, he said, which sought to attract people who would benefit from being close to the university and who would bring benefits to the university as well. Most park activities fell under one of four broad science and technology themes. One theme featured a new national center for weather and climate prediction that NOAA is building on 10 acres of land. The center would partner with the university and with NASA/Goddard Space Flight Center, and would hire about 800 employees. Already in the park are the Joint Global Climate Change Institute, run jointly by UM and the Pacific Northwest National Laboratory and funded by the Department of Energy. Another tenant is the Earth Systems Science Interdisciplinary Center, a $25 million-per-year program specializing on earth systems modeling.

The second park theme concerns language and national security in a broad context. A major initiative established under this theme involved both the Center for the Advanced Study of Language, a university-affiliated research center funded by the government, and a National Foreign Language Center that studied uncommon languages. This center has also established the Star Talk pilot immersion summer camp program. Two other programs on the campus that interfaced with this initiative were the National Flagship Language Program in Persian and the National Flagship Language Program in Arabic.

A third theme is food safety and agriculture and features four programs:

- U.S. Department of Agriculture Animal and Plant Health Inspection Services (APHIS).
- Food and Drug Administration (FDA) Center for Food Safety and Applied Nutrition (CFSAN).
- UM/FDA Joint Institute for Food Safety and Applied Nutrition (JIFSAN).
- UM Center for Food, Nutrition and Agriculture Policy.

Also in the park is the Animal and Plant Health Inspection Services, which partners with other activities on campus, including the Avian Influenza Project, a consortium of 17 institutions funded by USDA.

The fourth theme, intelligence, is focused around IARPA, the Intelligence Advanced Research Project Activity. This is a new government program that is modeled on DARPA. The objective of IARPA is to consolidate the highest level of forward-looking intelligence research. Dr. Mote said the recent recruitment of IARPA to the park was considered a major event, and one reason for the park's selection, he said, is the presence of other security-related activities—the Laboratory for Physical Sciences, which had already partnered with the National Security Agency for more than 50 years, and the Laboratory for Telecommunication Sciences. Other "seeds" for IARPA were the Joint Quantum Institute of NIST and the Center for Advanced Study of Language. The ability to partner with IARPA, said Dr. Mote, which would not be possible on a university campus, further illustrated the value of a research park.

He illustrated some physical features of M Square, which provides more than 2 million square feet of space on a 138-acre site. It has attracted some

$500 million of private construction and has received $5 million from the state, creating about 6,000 jobs. Important surrounding features include facilities of FDA, the American Center for Physics, Raytheon, USDA, the Metro/MARC Transit Station, and Baltimore-Washington International Airport. Less than a mile away and about eight miles outside the District of Columbia, a new mixed-use, 38-acre University Town Center (East Campus) is being created, allowing people to both live and work at the park.

He concluded by saying that although the park area is small in relation to many parks, it does make an important contribution to the university and the larger community. "The University of Maryland research park serves the mission of the university," he said, "by adding dimension to its partnership opportunities with industry and government on a global scale that cannot be fulfilled in any other manner that we have discovered."

DISCUSSION

In response to a question about the qualifications for inclusion in the park, Dr. Mote said that each proposal must be considered in light of four criteria:

1. What is innovative about this idea?
2. What is entrepreneurial about it—i.e., how are you going to leverage your money to get this done?
3. What kind of partnership is included that will expand your asset base?
4. To what degree is it an international project?

These criteria, he said, have acted as an effective filter for candidate projects.

Asked what kinds of policies could have made the park work better, Dr. Mote replied that once the state or county government sees that a park is credible and functional, more investment in the startup phase is key. He said that M Square is now probably beyond the "valley of death," where many new ideas perish for lack of support, but that the park has been helped by such "miracles" as the NOAA and IARPA agreements. Even so, timely financial support from the government would have reduced startup difficulties and accelerated its progress.

Asked what principles have led to the successful start of M Square, Dr. Mote cited the design of partnerships that bring benefits to both the university and tenants. He said that most tenants came because they were attracted by university assets. It would be even more desirable if it had additional land, with affordable housing and schools.

On the topic of potential conflicts between university standards and confidential or secret work, he said that graduate students were not allowed to participate as part of their degree programs or dissertations. They were allowed to work on such projects, however, and in fact their presence was attractive to employers and used as a recruitment tool. The university has also developed ways to hire people jointly—for example, faculty may have joint appointments at NIST and UM in nanotechnology. They offer special appointments for people at government labs, including the title of College Park Professor, giving no pay but otherwise conferring the rights of a faculty member, such as the ability to supervise graduate students' research projects.

In response to a question about intellectual property, Dr. Mote said there is significant effort to commercialize intellectual property developed at the university, led by the Office of Technology Commercialization, and UM-Net, a university-wide network of venture and commercialization services, researchers, and entrepreneurs. "They are constantly looking for entrepreneurs who want to do these kinds of things," he said, "and in the research park they can do proprietary work, with strong encouragement from us."

Leading Asian Models of S&T Parks

Moderator:
Lawrence Schuette
Office of Naval Research

Dr. Schuette asked the speakers in the first panel to address several basic questions: (1) What are we trying to do in regard to parks?; (2) What are the mechanisms and instruments to accomplish those goals?; (3) What level and type of funding is available?; (4) What achievements have been recorded; (5) What metrics are used to evaluate those achievements?; and (6) What are the current challenges?

He noted that the Office of Naval Research (ONR), a major sponsor of research around the world, is highly interested in the process of commercializing technologies. ONR sponsors research in 70 countries, has offices in four countries, including Singapore and Japan, and studies the issues associated with the valley of death. He said that in his view, "research parks appear to be an excellent place to cross that valley between the invention and the marketplace."

CHINA: NAVIGATING AT THE FRONTIER OF LIFE SCIENCES SILK ROAD

Zhu Shen
BioForesight

Dr. Zhu Shen is CEO of a California-based international consulting firm called BioForesight, which consults for companies in biomedical and related sectors. She began by saying that China is becoming one of the most significant world powers in science-based economic development. She said she would attempt to describe the "flavor" of the Chinese science and technology parks where she has the most experience: those in Beijing, Shanghai, and Suzhou.

She recalled the old Silk Road coming out of China as a key element of long-ago global commerce, and said that a modern version of the Silk Road exists. This time, China, instead of exporting spices, is "at the frontier of the Silk Road for the life sciences" and other technological innovation. The epicenters of this innovation, she said, are the major research parks, where many major research-based corporations are locating. She cited two of them as success stories. The first is WuXi Pharmatech (WX), which she said symbolizes the coming of age of the Chinese pharmaceutical industry. WuXi is the largest contract research organization in China, founded by her friend Dr. Guh Lee, one of the noted "sea turtles"—those who have gained their advanced education and work experience overseas and have returned home to become entrepreneurs and scientific leaders.

The second corporation is Hutchinson Medipharma, a UK-based company with R&D operations based in Shanghai's Zhangjiang High-Tech Park, which she called one of the world's most successful science and technology parks. She displayed a photo of signers of a strategic partnership between Hutchinson Medipharma and Eli Lilly, which she called a "very rich deal, like those we are accustomed to seeing in the U.S. between Western companies."

Growth of Pharmaceuticals in China

She said that the Chinese pharmaceutical market has "grown by leaps and bounds" in the past 15 years. In 1995 it ranked number ten in the world, has risen rapidly to number seven at present, and is predicted to become number five by 2010. "My prediction," she said, "is that it will pass the U.S. to become number one in the global pharmaceutical market in just two decades." The Chinese government, she said, has designated the pharmaceutical industry as a key to the growth of China. In addition to its increasing economic importance world-wide, the primary causes of death in China—cancer, cardiovascular disease, and heart disease—are similar to those in the United States and Europe so drugs developed locally will have global markets.

China's economic growth is not a new story, she said. Its GDP growth rate leads the world, averaging about 9.4 percent per year over the past three decades. The nation has grown from being the "world's factory," she said, featuring low-level jobs, to "the world's talent pool," with many S&T centers, outsourcing services, and new companies being formed in life science parks and high-tech parks. Today China has more than 54 state-level economic and technological development zones, and 53 national high-tech development zones.

The Beijing Park

She illustrated the dramatic pace of change in China with pictures of what is now Zhongguancun Science Park in Beijing, both 20 years ago and today. "This is where I grew up," she said. "The changes have been transformational." The park hosts over 20,000 enterprises and 950,000 employees, receiving total income of 850 billion Yuan (about US$ 110 billion). More than 800 enterprises have income exceeding 100 million Yuan.

Of the industries represented in the park, the majority (56.6 percent) are classified as information technology, 12.5 percent as "new energy," 12.3 percent as biomedicine, 9.4 percent as advanced manufacturing, and 8.4 percent as new materials. The park has attracted almost 10,000 "sea turtles," she said, who have set up 4,200 companies in Zhongguancun Science Park.

The Shanghai Park

She turned to Shanghai's Zhangjiang Biotech and Pharmaceutical Base, which is part of the Zhangjiang High-Tech Park. This is considered by the Chinese leadership and industry insiders to be the most successful biotech and pharmaceutical park in China. Started in 1992, Zhangjiang High-Tech Park has developed 17 out of 25 square kilometers and hosts more than 3,600 companies, of which more than 140 are foreign. It employs more than 100,000 researchers and other workers.

The Zhangjiang High-Tech Park emphasizes three major areas: life science, which accounts for about 50 percent of revenues; software; and information technology. Its corporate tenants in the life sciences include six of the world's top ten pharmaceuticals: Roche, Eli Lilly, Pfizer, Novartis, GE, and AstraZeneca. IT tenants include Hewlett-Packard, Lenovo, Infineon, and Intel. Software tenants include IBM, Citibank, Infosys, SAP, and eBay. Specialty chemical companies include DSM, Henkel, Solvay, Dow, DuPont, and Rohm & Haas. Among the biotech tenants are not only foreign companies, but more than 110 indigenous firms, she said, "many of which I know personally." These include more than 60 small-molecule drug development companies, 35 medical device and diagnostics firms, and more than 15 traditional Chinese medicine companies. Hutchinson

MediPharma began as a traditional medicine company, but leadership quickly built up its novel drug discovery capability, and this attracted Eli Lilly.

The Zhangjiang High-Tech Park is located in Pudong, a district of Shanghai. Started in 1992 on undeveloped farmland, Pudong has emerged as a major financial and commercial hub. Dr. Shen noted that the 228 square kilometer Pudong site accounts for 25 percent of Shanghai's GDP, 50 percent of foreign trade, and 30 percent of foreign investment. It is also "a pilot area for reform." "I don't think we can find a park like this in the United States, yet," she said with a smile.

In line with Shanghai' new slogan, "Better City, Better Life," the Zhangjiang High-Tech Park organizers recognize that "the economy is not everything." They want to represent Shanghai as "a city that is truly metropolitan and international, with an open-minded population, clean environment, and beautiful parks that can attract returnees and talent from the world over."

The Suzhou Park

The third Chinese park she described is the "up-and-coming" Suzhou Industrial Park (SIP). Established in 1994 in a location "well known for its classical gardens and beautiful ladies," she said. Suzhou today is known for its "innovative spirit and ability to attract top talent world-wide." It is a joint development between the Chinese and Singapore governments, which she called unprecedented, it is considered first among all regions in China in "pro-business mentality," efficiency, and consistency of policies. Located 80 kilometers west of Shanghai, Suzhou has taken its place at the high-tech frontier of the global economy, she said. In land area only 0.1 percent and in population 0.5 percent of China, it accounts for 2.3 percent of GDP, 1.5 percent of financial revenue, 10 percent of imports and exports, and 8.3 percent of foreign investment. Of the Fortune 500 companies, 113 have set up operations in Suzhou.

Some Lessons from the Chinese Experience

Dr. Shen said that even though the Chinese S&T parks had been started only two decades ago, long after the major initiatives in the United States, there were valuable lessons to be learned:

- Government support is critical. The Chinese government has invested more than US$ 1.4 billion in the Suzhou park alone.
    - Governments must be involved at the national, regional, and municipal levels.
    - The right policies, incentives (tax waivers, free rent), infrastructure, financing, and resources are needed to attract multinationals, "sea turtles," and local entrepreneurs.

o Park firms need many business resources: one-stop services such as accounting, IP advice, corporate counseling, and VC investment.

- Competition creates competence, and "if your park isn't good enough, the top talent won't come."
- Cooperation fosters excellence, especially cooperation between home-grown companies and those of the government research institutes.
- Visiting S&T parks worldwide can bring valuable lessons.
- The arrival of "sea turtles" generates change.

Dr. Shen is also associated with SABPA, the Sino-American Biomedical and Pharmaceutical Professional Association. Established in 2002 in San Diego, it has more than 1,500 members, one-third of them non-Chinese. Its mission is to stimulate networking, career development, education, and Pacific alliances, and it highlights its presence through an Annual Pacific Forum held in November each year in San Diego. She also summarized the activities of BioForesight, which include corporate developing, financing assistance, corporate strategizing, and media relations.

She concluded by suggesting that the Apple iPod is a possible model for the global alliances of the future. Designed in the United States and manufactured in China with technology from India, it is truly a global product—but Apple captures most of the profit from sales of the iPod. This is a model for the science parks and indigenous companies in China to ponder, she said—how to control the major value chain of their products.

THE SINGAPORE SCIENCE AND TECHNOLOGY PARK

Yena Lim
Singapore Agency for Science, Technology and Research

Ms. Lim gave a brief introduction to Singapore, which she characterized as an island state with an area slightly smaller than that of New York City, and one of limited natural resources. Since winning independence from Malaysia in 1965, the population has approximately tripled, from 1.6 million to 4.5 million, while its economy has grown a hundredfold, from S$2.1 billion to S$210 billion. During this period, she suggested, Singapore moved quickly through major stages of economic development—from labor-intensive in the 1960s and 1970s, to skill-intensive in the 1980s, to technology-intensive in the 1990s. GDP per capita has grown from a mere US$512 in 1965 to US$35,640 in 2006—very close to the U.S. GDP per capita of US$47,330. The education level of the population had improved significantly with 24 percent of the population entering university in 2006 and 41 percent entering polytechnics.

Tradition of Excellence in Math and Science

She attributed this rapid progress at least partly to Singapore's "ability to reinvent and transform itself." It watched closely as the world began to understand the capabilities of China and India and how these high-growth nations would change the economic order. The Singapore government decided to leverage on its tradition of excellence in science, mathematics, and technology and embarked on an ambitious growth strategy that invested substantial resources in R&D to catalyze knowledge and innovation-intensive economic activities. The strategy has clearly paid off. According to the World Competitiveness Yearbook 2007, Singapore ranked second in competitiveness, after only the United States. It ranked first in "Ease of Doing Business," according to the World Bank's "Doing Business 2008," and seventh on the Global Competitiveness Index.

In mathematics and science, Singapore led the world in the TIMSS report for 2003. Singapore students have done very well internationally, ranking first in Grade 4 and Grade 8 mathematics, and first in Grade 4 and 8 sciences. "Students also have more interest in pursuing tertiary education in the sciences," she said, "as they see government's strong commitment to creating an environment that is conducive for science and technology. These open up many employment opportunities for the long term."

In terms of economic structure, manufacturing is the strongest sector in Singapore, accounting for 28 percent of GDP in 2006. Within manufacturing, the lead sectors are chemicals (33 percent) and electronics (32 percent). But "the interesting story for the last few years," said Ms. Lim, "has been the knowledge-intensive biomedical sector, which has grown to 10 percent of all manufacturing output because of the inflow of top-ranked talent, strong base of biomedical research capabilities and state-of-the-art infrastructure."

In response to government policy, the economy is becoming more R&D intensive. In 2006, Gross Expenditure on R&D (GERD) was 2.4 percent of GDP, and the target is to reach 3 percent by 2010. Of this, about two-thirds of R&D expenditure was from the private sector. The flourishing research scene has led to greater research efficiency. At a Wiley-Blackwell Research Seminar in March 2008, Blackwell's bibliometrics director Iain Craig provided data illustrating that the research output of Singapore was on track to reach and then exceed the world average in the next few years, having increased by some 72 percent from 2000 to 2007. With an R&D expenditure of US$3.1 billion in 2006, Singapore generated publications at a rate of 0.3 publications per researcher, higher than that of either China or Japan.

Ms. Lim described the national R&D framework in broad terms, saying that the government has allocated a 5-year budget of S$13.55 billion (nearly US$10 billion[4]) for R&D. Of this, the National Research Foundation has S$5 billion, and

---

[4]In comparison, the FY2009 budget request of the U.S. National Science Foundation is US$6.85 billion.

invests primarily in long-term projects and is overseen by a Research, Innovation and Enterprise Council chaired by the Prime Minister. "The top-level commitment to investing in science and technology for the long term," she said, "will ensure that the Singapore economy has the capability to re-invent itself and keep its international competitiveness."

The budget of the Ministry of Education is about S$1.05 billion, and the Ministry of Trade and Industry receives about S$7.5 billion to promote R&D in the public and private sectors. Of the latter amount, S$5.4 billion goes to A*STAR, the Agency for Science, Technology, and Research. A*STAR funds the Biomedical Research Council (S$2.4 billion) to "deepen basic science capabilities and promote translational and clinical research"; the Science and Engineering Research Council (S$2.4 billion) to build multidisciplinary research in support of industry; the A*STAR Graduate Academy (S$450 million) to develop future research scientists and engineers; and Exploit Technologies (S$133 million) to commercialize the intellectual property of A*STAR.

Biopolis: A Biomedical Hub

She turned then to the Biopolis project, planned as "the biomedical hub of Asia," a city within a city intended for scientists, researchers, and entrepreneurs. Located in central Singapore, it is designed to attract scientists from all over the world who will come for the quality of scientific research and the cosmopolitan work environment. "At the end of the day," said Ms. Lim, "it is the quality of science and the concentration of top-notch researchers working together that will have impact." In addition to the policy goal of encouraging public-private partnerships, the buildings themselves are intentionally compact and close together, so that "when the researchers come out of the lab into a common space, they have no choice but to interact with one another."

She said that the planners of Biopolis asked themselves what it would take to create an environment that will attract not only top talent, but also people who have families. They decided against a "normal" science park, one that went dark each evening after office hours. Instead, the plan was to develop a self-sufficient S&T hub with areas dedicated to specific research fields; open and connective structures that would encourage interaction and innovation; and facilities for comfort and relaxation. The design brief, she said, was to be human-centric, to make the facilities both convenient and stimulating.

Designed to Promote Interaction

Both small and large details were designed to bring people together. For example, the buildings are connected by sky bridges so people can walk from one research institute to another—and meet each other as they do so—without having to descend to ground level or drive. For further interaction, Biopolis has an auditorium, lecture theatres, and meeting rooms. In and around the laboratory buildings are convenience stores, a gym, a childcare center, restaurants and cafes, and a pub.

For the work environment, many facilities have been designed to be shared by scientists and engineers in related disciplines, including the Zebra fish facility, bioreactor, electron microscopy, proteomics, MRI, histology, x-ray crystallography, DNA sequencing, flow cytometry, lab supplies, media preparation, and glassware washing.

Biopolis was conceived and built quickly, said Ms. Lim. Phase I groundbreaking happened in December 2001, and Biopolis was officially opened in October 2003. Biopolis Phase 1 has an area of 185,000 hectares, Phase 2 has 37,000 hectares, and Phase 3 (scheduled for completion in 2009) will provide another 41,500 hectares. Future phases are on the drawing board. The buildings, under the aegis of A*STAR@Biopolis, have names indicating the kinds of activities supported. For example, Genome houses the Genome Institute of Singapore; Nanos has the Institute of Bioengineering and Nanotechnology; and Proteos houses the Institute of Molecular and Cell Biology. Major companies such as the Novartis Institute for Tropical Diseases and the GlaxoSmithKline Center for Research in Cognitive and Neurodegenerative Disorders have built a presence in the Biopolis.

Fusionopolis: The Physical Sciences and Engineering

Hot on the heels of the Biopolis is Fusionopolis, located about half a mile from the Biopolis. Fusionopolis is designed to be an integrated and comprehensive work-live-and-play environment. Slated to open in October 2008, it will have public- and private-sector labs, homes, service apartments, hotels, a shopping mall featuring smart-shopping technologies, food and beverage outlets, and an experimental theatre for art performances.

Envisaged to be a one-stop science and engineering R&D powerhouse, it will offer a broad array of capabilities in an integrated manner, in the fields of physical sciences and engineering, to address complex problems facing industry and society. It will focus on the physical sciences and engineering research, especially:

- Energy (fuel cells, organic photovoltaics).
- Home 2015 (assistive technologies for health monitoring and rehabilitation, medical devices and technology).
- Aerospace (computational fluid dynamics, manufacturing processes, automation).
- Nanotechnology.
- Sensors and sensor networks (wireless communication and robotics for remote monitoring).
- Cognitive science, with future applications in social robotics.

The watchword for Fusionopolis, she said, is integration: of R&D capabilities in the physical sciences and engineering; of the physical sciences and engineering with biomedical sciences; and of academic and industrial science, to speed research results along the path toward economic application. A*STAR will also play a role in the last phase, integrating the work of research institutes with MNCs, SMEs, and start-ups, as well as with other agencies, such as the Economic Development Board and SPRING Singapore.

Finally, both the Biopolis and the Fusionopolis are located within One-North, which is a larger area situated within a science and education talent belt, that also encompasses the National University of Singapore, the National University Hospital, part of the Nanyang Technological University, Singapore Science Park, and the Ministry of Education. There are public housing estates nearby, and there are plans to build a major hotel and a large number of condominium units. The research community consists of a community of more than 3,000 research scientists and engineers from more than 50 countries, bringing a mix of talents and including a major commitment to the arts and art creation in situ. One-North is well-located—just 10 minutes from the city center and 20 minutes from the Changi International Airport. It is planned, she concluded, as a vibrant dynamic environment because "the research community must not be isolated." The ambitious premise of the design is to create "an ecosystem designed to nurture new ideas and push them quickly to reality."

INDIAN SCIENCE AND TECHNOLOGY PARKS

M. S. Ananth
Indian Institute of Technology-Madras

Professor Ananth began by giving some recent context for research park development in India. Before the early 1990s, the national government was strongly socialistic and kept the country's economy largely closed. Government leaders did not encourage the kinds of government-industry-academic partnerships that characterize modern S&T parks. As a result, no formal research parks were located near universities, although some inter-sector partnerships were encouraged informally within campuses.

In 1990-1991, however, the government of India launched a process of economic liberalization. Discussions about parks began, and some university-industry partnerships began to form and move off campus into formal research parks. "When I went to the government and asked them for formal permission to build research parks," recalled Dr. Ananth, "I quoted Louis Pasteur, who said, 'Discovery is the result of chance meeting of prepared minds.' I said that I have been preparing minds for such opportunities for 50 years." He explained that chance had been meeting prepared minds in the United States for many years, and that all the discoveries were occurring there. "We needed to wake up to the fact that we must create these opportunities in India as well," he told the government.

Beginnings of the Park Movement

By about 1999, the S&T park movement began to take off, he said, as India's entrepreneurial spirit was being liberated. Initially, India took advantage of its strong cost advantage, but this advantage is disappearing as India moves up the value chain.

The backbone of Indian higher education in science and engineering is formed by its 12 S&T institutes of national importance. These include the seven Indian Institutes of Technology, five of which were formed soon after independence in the 1950s (including Dr. Ananth's institute in Chennai); one was added in 1995 and another in 2001.[5] The strong national recognition and status of the IITs makes them logical anchors for research parks as they provide leadership in every field of science and engineering.

The objective of the Indian science and technology parks, said Dr. Ananth, is to promote and foster the spirit of innovation. The nation's parks still have a long way to go in terms of the infrastructure and support systems necessary for competitive R&D, he said. Nonetheless, economic growth has been remarkable—consistently around 9 percent, he said, with manufacturing growing at 12 percent.

The parks are intended to:

- Incubate early-stage entrepreneurial ventures based on technology and innovation.
- Facilitate networking with professional resources for the incubated companies.
- Identify technologies and innovations that have potential to be commercial ventures.

Indian research parks are still relatively small and not generally associated with universities. The bigger ones have varying mixes of tenants and partnerships. One group of parks is the Andhra Pradesh Biotech Parks. These include:

- Shapoorji Pallonji (SP) Biotech Park near Hyderabad. It has about 140 acres under development and contains about 17 companies with an investment of about Rs 4 billion (or approximately US$93 million).
- The Marine Biotech Park occupies 218 acres near Visakhapatnam. In association with the Andhra University, it focuses on marine resources, marine foods, nutraceuticals, and fisheries.

[5] Additional IITs will be inaugurated in 2008-2009.

• The ICICI Knowledge Park is focused on facilitating business-driven R&D. It is located on 200 acres of land near Hyderabad and holds 13 companies with about Rs 420 million (or approximately US$9.8 million) invested.

The state of Andhra Pradesh also has an Agro Park on 200 acres in the International Crops Research Institute for the Semi-Arid Tropics campus. The park comprises an Agri-Biotech Park, an Agri-Business Incubator, a Hybrid Seeds Consortium, and SAT Ecoventure. Three ventures have been developed in the Agri-Biotech Park, among them a facility for testing aflatoxin contamination in food crops.

Elsewhere in the country, the Maharashtra Biotech Park is an international park, a joint venture between the Maharashtra Industrial Development Corporation and TCG Urban Infrastructure Holding Ltd., a Chatterjee Group company. The total investment is Rs 2.5 billion (or approximately US$58 million) in a 110-acre park near Pune, which now holds three companies.

The Uttar Pradesh Biotech Park has been set up in collaboration with the Council of Scientific and Industrial Research, universities, and industries in Lucknow, the capital of Uttar Pradesh. Here there are several biotech firms, biotech consulting services, clinical research centers, and other health-related companies, as well as the Software Technology Park of India and BioAlliance of Germany.

In Tamil Nadu, where Dr. Ananth lives, Ticel Bio-Park has been developed in 2004 by the Tamil Nadu Industrial Development Corporation on a five-acre site at a cost of Rs 625 million (or approximately US$14.5 million), in collaboration with Cornell University; it now has three occupants. Tamil Nadu also has the Golden Jubilee Biotech Park for Women (2001), which Dr. Ananth believes is the first Asian biotech park dedicated exclusively to furthering the careers of women entrepreneurs. Covering 20 acres and holding 13 tenants, Golden Jubilee has created a database of over 500 technologies or projects covering about 150 research institutions.

Partnerships with the IITs

While these parks have been set up without major university affiliations, others have formed partnerships with the IITs and the Indian Institute of Science. Dr. Ananth noted that this is to be expected, since "entrepreneurship flourishes in the vicinity of high-quality educational institutions." In addition, he said, "research parks spark innovation through the 'idea-factory' approach by bringing together three kinds of minds and experience: faculty, with knowledge of fundamentals; students, with their spirit to conquer the world; and S&T personnel, with their ability to convert ideas into marketable products. In America you may take this for granted," he emphasized, "but you have to create this in other countries." Dr. Ananth listed the following university-initiated projects:

- One of the earliest university-affiliated parks is the Society for Innovation and Development, with a campus of about 400 acres, in partnership with IISc-Bangalore. Started in 1991, it has attracted participating companies in bioscience, computer services, software, pharmaceuticals, tropical diseases, and biosciences, as well as Tata Motors.
- A newer research park is the Society for Innovation and Entrepreneurship, set up in 2004 on the campus of IIT-Bombay (located in Mumbai) as a business incubator. It now includes facilities covering 10,000 square feet and is supported by the Department of Science and Technology, the Technology Development Board, the National Entrepreneurship Networks, the IIT-Bombay alumni, and the Ministry of Communication and Information Technology.
- The Foundation for Innovation and Technology Transfer (FITT) is an industry interface of IIT-Delhi established on the campus in 1992. It has been conceived and implemented as a technology business incubator by IIT-Delhi to provide infrastructure and techno-managerial support; funding source data on bridge capital for start-up companies; and networking with venture capital companies.
- The IIT-Kanpur, in partnership with the Small Industries Development Bank of India (SIDBI), started an Innovation and Incubation Centre to scale up laboratory-proven concepts to commercial scale. Located on the IIT-Kanpur campus, it has three "tiers": a "nursery" incubation project, technology-based start-up companies and technology/R&D units of an existing SME desiring close technology interface with IIT-Kanpur
- IIT-Madras Research Park, located in Chennai, is an independent company promoted by IIT-Madras with support from the government and alumni. It is outside the campus, he said, "but within cycling distance." It is planned to have an area of 1.2 million square feet, built in three phases, to house R&D activities of companies wanting to work with IIT-Madras as well as companies to be incubated. Some 27 companies have signed up, a few of which are MNCs. Many of them are automotive firms, including Tata, and automotive parts companies; Chennai is a center of automotive research in India. The goal is to have some 5,000 development engineers and finishing schools for up to 5,000 students per year.

Dr. Ananth elaborated on the plans for the IIT-Madras Park, in which his colleague Dr. Ashok Jhunjhunwala and he have played a significant role. He noted especially the synergies to be gained by leveraging university expertise and knowledge, and by offering higher education opportunities for the park's employees. Some 5 percent of R&D personnel are expected to teach as adjunct or permanent faculty, while 10 percent are expected to register for part-time masters and PhD programs in IIT-Madras.

To the extent that this and other parks can attract highly qualified individuals to R&D, he said, India could become a "design house" that develops higher

FIGURE 1 The proposed IIT Madras Research Park.
NOTE: The first tower will be inaugurated by the end of 2008.

quality products, participates in international standards bodies, and develops intellectual property rights. To advance such goals, the Chennai Park will feature infrastructure for start-ups, availability of venture capital, consultancy, and prototype firms in the vicinity. Some 15 percent of park space will be reserved for start-ups and for facility and training companies at concessional or deferred rental. The remaining 85 percent of the space will be rented or leased at long term to companies leveraging R&D partnerships with the university.

Earning Credits to Remain in a Park

To remain in the park, each company must earn a minimum number of credits by interacting with IIT-M. The credit system is designed to promote entrepreneurial activity, inter-sector interaction, and partnerships. Thus, companies are given credits for the following actions:

- R&D projects with IIT-M.
- Consultancy to IIT faculty.

- Earning royalties.
- Sponsored PhD/masters students.
- Serving as adjunct faculty.
- Teaching at IIT-M.
- Mentoring PhD/MS/BTech students.
- Giving part-time employment to PhD/MS/Btech students.

The total cost of the park has been modest, said Dr. Ananth, at approximately 3 billion Rupees. This money has come mainly from the government and from bank loans, while some donations from IIT-M alumni and internal accruals will make up the balance. The local government has also provided 11.5 acres of land and infrastructure that includes an approach road and electricity. It has made economic concessions to the park, including tariff concessions for electricity, sales tax, excise and custom duty, and floor space index. The park enjoys an interest-free, long-term loan from the central government of about Rs 1 billion (or approximately US$23 million).

Dr. Ananth concluded by emphasizing the importance of incentives that promote university-industry interaction. "Traditionally," he said, "we have found that one of the problems in situating parks near campuses is that they focus only on the real estate and they don't interact with the university. We are working hard to change that."

DISCUSSANT

Phillip H. Phan
Rensselaer Polytechnic Institute

Dr. Phan said he would make some specific observations about the science and technology parks in Asia and some general comments about science park policy.

His first observation, relevant to both federal and state governments, was that innovation and entrepreneurship must become incorporated not only as they pertain to economic development policies, but also as cultural values. He said that India's long history with the IITs has raised the profile of science and technology and created "a natural base on which many of its park initiatives are built." In Singapore, the national government had built powerful educational programs from kindergarten through postgraduate levels, raising the value of learning and rewarding scientific and engineering excellence. "Without this fundamental element," said Dr. Phan, "all the real estate and all the parks you build won't go very far. You can't build an innovative society purely from capital investments. That's where you start, but eventually you have to build an indigenous ability to innovate."

Importance of the Right Metrics

Dr. Phan emphasized the importance of finding the right metrics to evaluate S&T parks. Because parks are expected to be vehicles for innovation, the metrics will be different from those used to evaluate scholarship or business. He said that many park planners, especially in Asia, regard a research park primarily as a vehicle for attracting foreign direct investment and rate parks in terms of how many high-tech people they attract. "You don't build a park just to generate employment," said Dr. Phan. "You plan to create the conditions and lifestyle that will attract innovative people and the resulting innovation. You can't just 'build it and they will come'—you have to make it attractive in the first place."

Another feature that characterizes the more successful park development efforts, he said, is access to risk capital. Numerous studies, he said, demonstrate that local access to risk capital must be a precondition rather than an outcome of park development.[6] Enabling access to exit markets, so risk capital providers can cash out, he added, should also be a policy target. If government-subsidized programs are to create strong business opportunities, firms must be able to move capital in and out quickly, and to rely on prompt returns on investments.

Drawing on the Best Qualities of Public and Private

He emphasized the need to create real partnerships between state-owned companies and private firms. Despite their different cultures, these sectors benefit from the park only when cooperation is deliberately embedded into their relationship. This, he said, can create a desirable kind of semi-public, privately owned entity that brings to bear the best qualities of both to promote innovation.

He noted that in Asia it is possible and often necessary to offer special perks and a living environment that attract employees to parks, a lot of which goes on behind the scenes during the planning process. Some of these conditions may be harder to create in the United States, such as special economic export zones with tax and currency controls, and may be more realistic for local governments rather than the federal government to implement. In Asia, education is a critical component of innovation policy, especially the teaching of English. He said that China is trying to create a "parallel," private education system to accelerate the learning of English.

He mentioned two final challenges for Asian parks. One is a fundamental cultural gap concerning the researchers themselves. Most parks depend on proximity to universities, but university scientists do not traditionally see entrepreneurship as a priority. Another is the need for a supportive legal framework that addresses

[6]For a review of recent studies, see Phillip H. Phan and Donald S. Siegel, "The Effectiveness of University Technology Transfer: Lessons Learned from Qualitative and Quantitative Research in the U.S. and U.K," Rensselaer Working Papers in Economics 0609, Troy, NY: Rensselaer Polytechnic Institute, Department of Economics, 2006.

intellectual property issues. "If anything is going to be a problem for China," he said, "it is going to be dealing with intellectual property."

Policy Implications for the United States

He described several federal policy implications for the United States. He proposed first that the federal government focus primarily on creating conducive legal and political institutions. He warned against overly restrictive intellectual property protection that can stifle creativity and innovation, citing the findings of Lawrence Lessig, chair of the Creative Commons project.[7] He also recommended that the government consider unrestricted immigration for scientific, engineering, and creative talent. Every country building up S&T capacity is doing everything possible to import talent, and the United States will face a disadvantage unless it does the same. "The growth of endogenous talent can never exceed the total demand for talent," he said, "so there is always a need for more. If you restrict flow of talent from outside, you are diminishing your absorptive capacity to exploit existing talent."

He concluded with several recommendations for measures to support parks. In terms of state and local policy, he said, states should recognize that straight subsidies do not work very well. Any subsidies should accomplish some kind of leverage, such as education policies or incentives to attract risk capital from foreign sources. He advised against plain subsidies simply to compete with neighboring sites or regions in favor of investments in math, science, and liberal arts education, especially at the K-12 level; incentives for foreign and domestic college students to enroll in science and engineering programs; and reduced financial costs (e.g., business taxes and rules) for young foreign and domestic knowledge-based enterprises.

DISCUSSION

Dr. Shen said that in China a good deal of private money is waiting to "piggyback" on the activities of high-tech parks—so much so that parks are becoming more selective about tenants. There are committees, she said, to evaluate prospective tenants from the private sector according to a set of rigorous standards.

Ms. Lim noted that prospective tenants do not get any special support for joining a park. On the contrary, they are eager to come "mainly because of the special environment that we and other tenants have created there."

Dr. Phan reiterated his opinion that subsidizing park activities with infusions of capital tends to distort the market. Unless local government subsidies have sunset provisions, he said, subsidies simply transfer money from public to

[7]Lawrence Lessig is a professor of law at Stanford Law School, founder of the school's Center for Internet and Society, and author of numerous books.

private entities. "The real issue," he said, "is that you want these companies to want to locate in the parks. The key is to have policies that make it possible for risk capital to follow."

Dr. Shen commented that in China there are clusters of research institutes, universities, and laboratories around all the prominent high-tech parks. "The idea is that these are places where a lot of the top talents from different fields are clustered—this then is what attracts private enterprises. There is a multilateral collaboration among government-sponsored research institutions and those of the private industries, both locally and internationally."

Dr. Wessner ended the panel discussion by reminding the participants that "if the markets were working fine, none of you would be here. Markets are rarely, if ever, perfect, and you're trying to make them work better." In particular, he said, the markets do not work well in providing early-stage finance for start-ups. "Planners in Asia have looked at the outcomes of two centuries of markets that didn't work to the advantage of young companies, and decided that they would like to change the terms of those markets. That's what they are doing, just as the United States did 50 and 100 years ago. The debate is not one of whether something called 'the market' is 'efficient,' or whether it is 'distorted.' The issue, more practically, is how to help real-world markets work better by applying the right policies."

North American and European S&T Parks

Moderator:
Peter Engardio
BusinessWeek

Mr. Engardio continued the discussion by noting that development of S&T parks is "a global game," and that the transition from the parks in Asia to those in North America and Europe would build a comprehensive picture of how planners around the world are adopting similar policies to spur innovation.

THE ENGLISH EXPERIENCE

Jane Davies
Manchester Science Park
United Kingdom

Ms. Davies said she has been CEO of Manchester Science Park Ltd. (MSP) since 2000. She also chairs the U.K. Science Park Association (U.K.SPA), the umbrella membership organization, which has 67 member parks and 10 in the planning stages. Together they provide nearly 2 million square meters of accommodation, space for 3,400 companies, and about 75,000 employees. She noted that things in the "old country" are a bit smaller than recently described projects in Asia.

She said that parks in the U.K. also differed from U.S. parks in both scale and ownership. The early science parks in the U.K., built in the 1980s, tend to be owned by universities and operated as income-generating properties. However, since technology transfer and business incubation have emerged as important drivers in the development of modern economies, the public sector has also become active in stimulating and funding new parks; Manchester is one example of such a shift. In the 1990s and in the 21st century, parks have been developed in the U.K. and in Europe with capital funding from regional development agencies which see science parks as tangible evidence of their region's developing knowledge economy.

A Mix of Ownership

Many of these parks, she said, are partnerships between government and local universities, which in the U.K. receive funding from central government for this "third-mission" activity. She likened it to the development described earlier for the University of Maryland. More recently, private property developers have taken an interest in the science park idea, beginning to forge their own partnerships with public entities and adding yet more kinds of collaborations to the mix. Today about 11 percent of parks are privately owned, 19 percent are university-public partnerships, 27 percent are university-owned, and 43 percent are partnerships between universities and other public or private bodies.

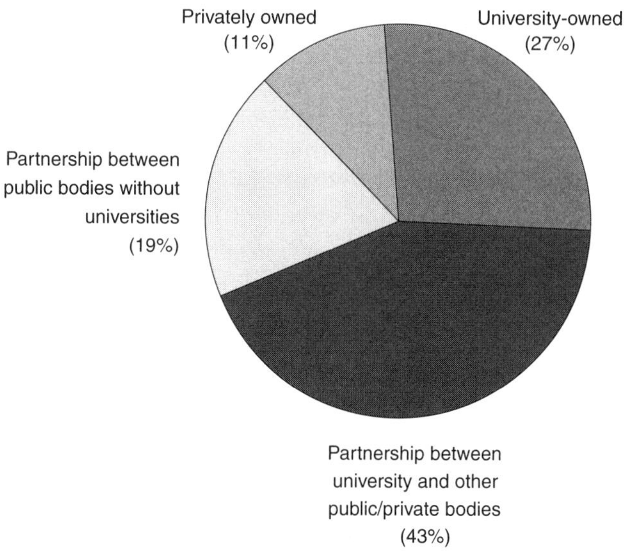

FIGURE 2 UK parks ownership profile.

In 2003, the U.K.SPA contracted with Angle Research to examine the impact of science parks on the U.K. economy. They examined some 900 companies altogether, comparing the economic and innovation performance of park tenants with similar firms located outside parks. The results showed that the single most important factor affecting the performance of science parks is the state of the sub-regional economy in which they operate. Also, the companies in parks were found to have higher growth rates, in terms of both turnover and employment, and better access to risk financing.

Manchester Science Park

She turned to Manchester Science Park itself, which is one of the older S&T parks in the U.K., started in 1984 as a 22,500-square-foot building on a 15.5-acre site. Today there are 305,000 square feet of buildings on three sites that employ about 1,100 people. She offered some of the reasons for its success, beginning with its location next to the campus of Manchester University, a major research university which aims to become world-class in science and technology. Another advantage is a corporate structure that guarantees every tenant flexibility and independence. The city has strong civic leadership with a vision of becoming global. It has a history of working in partnership with the private sector and is the only U.K. city outside London that has "the economic assets and the will" to become the second engine of the U.K. economy. Even local strength in sports can be a recruiting asset for the park, which proudly reminds candidates of the presence of Manchester United, one of the U.K.'s premier football clubs.

Like so many parks, MSP was created as an economic development initiative. Manchester was badly weakened by the downturn in manufacturing of the 1980s, suffering job losses and much personal pain. Having heard about and visited Research Triangle Park, representatives of Manchester government, the university, and the commercial sector came together to set up a science park, raising an initial capital investment of 210,000 Pounds. The city built the first building, and MSP was engaged to manage it. In 1999, MSP formed a joint venture with a regional builder-developer, Pochin Plc. This venture successfully developed three buildings on a site adjacent to the main campus, one of which was sold in 2005.

The park has received only limited public funding in its 24-year history, in the form of gap funding for three of the buildings. All the partnership activities of tenants and the university have been paid for out of MSP's profits. The park has never paid a dividend to investors, who are content to see the value of their holdings increase as a result of park activities.

Measuring Success

In measuring success, the park uses the strategic objectives of economic development and knowledge exchange. The first metric is growth in tenant companies. "We are much more interested in the companies than we are in the park itself," said Ms. Davies. "We look at how our companies are doing as a measure of the impact we're after." Tenant companies are asked to fill out a questionnaire annually, and growth is measured by employment, company turnover, and expansion within the park.

To measure innovation, companies are asked how they are developing their links with higher education, such as how many graduate students they employ. They are asked about their licensing activity and about new products and services; and about sources of investment (friends and family? venture capital? stock offerings?) which indicate how successful they are at bringing new money into the local economy.

The park management also provides assistance to its companies and even tracks the alumni companies to monitor their development. In 2007, for example, MSP found that 79 percent of the companies operating in 2001 were still in business. By comparison, the average survival rate of all firms in Manchester is 64 percent. MSP also found that 70 percent of companies that left the park were still operating in the city region. For Manchester, said Ms. Davies, this figure is just as important as the growth of the park companies.

The next science parks to be built in the U.K. will represent a third generation, she said, and each is likely to be:

- A global player with local roots.
- Part of the local community.
- A healthy business (making money and attracting investment).
- An essential element of university activity.
- Part of a multiplicity of networks.
- Focused on the needs of tenants.

The Challenge of Small Size

In reviewing some challenges for MSP, Ms. Davies said that the most important one is its small size. "We are making a positive impact," she said, "but because of our size we cannot make a regional impact as we would like to do. We're looking at expansion in the city region. I think we'll see in the U.K. that some parks will come together under common management." She cited an example from the Basque country of northern Spain, where three parks merged under one board so as to contribute more effectively to the region's prosperity.

Another challenge locally is the patchy nature of the innovation system in the Manchester city region. Further, finding sufficient risk capital remained a challenge, as did the problem of young people avoiding science and engineering. "We need ways to rebuild our resources of bright people," she concluded, "by showing students that science and engineering offer exciting careers. I think if we started paying them as much as bankers, that would have a bigger effect than any program at kindergarten level."

MONTERREY: INTERNATIONAL CITY OF KNOWLEDGE PROGRAM

Jaime Parada
Research and Innovation Technology Park (PIIT)

A new S&T park is unfolding in Monterrey, Mexico, said Mr. Parada, with much support from the federal government. The city of Monterrey has an advantageous location near two U.S. border crossing points and 320 miles from the major east-coast port of Altamira. The city also has a dynamic industrial base, 4.3 million inhabitants (4 percent of the population), a GDP of US$69.2 billion, and per capita income of US$15,975, nearly twice the national average. The city produces 11 percent of Mexico's manufacturing goods, equivalent to US$12.1 billion.

The province of Nuevo Leon also has a good higher educational system to anchor park activities, including 93 colleges and universities. Monterrey Tech and the University of Monterrey, he said, are two of the most prestigious universities in Latin America, and the University of Nuevo Leon is ranked as one of the best state universities in Mexico. The province also has a strong base for R&D, with more than 1,500 researchers in public and private research institutions.

Preparing for Long-term Alliances

The core ingredient in the Monterrey park strategy, he said, is to prepare for long-term alliances among universities, businesses, and government. The park will be oriented to achieve economic growth and quality of life through the "triple helix" of education, R&D, and innovation.

He summarized the park's long-term (2025) vision as expecting:

- To increase the state's GDP per capita from $15,975 to $35,000 by 2020.
- To become one of the world's 25 most competitive regions.
- To consolidate a world-class education, research, and innovation system.

• To demonstrate to the regional population the importance of education, knowledge, R&D, and innovation in their lives.

He acknowledged that "Monterrey, City of Knowledge" is "a grand name," and suggested seven strategies designed to live up to it:

• To redesign the curricula of the education system in Nuevo Leon to incorporate S&T and innovation.
• To enhance the existing universities and R&D centers and attract new research centers and investment in technology-based companies.
• To promote innovation in existing companies through tax incentives and other measures.
• To create new innovation-based companies using appropriate financial instruments.
• To promote a new culture of innovation in Nuevo Leon society.
• To create a competitive urban infrastructure and quality of life.
• To generate the necessary legal framework, governance, programs, strategic projects, and mechanisms to sustain park development for 25 years.

Monterrey site planners have studied many parks around the world, but Mr. Parada acknowledged that building the PIIT is "a brand-new experience for us." The site has the potential advantage of two anchors, one being the airport located five minutes away, and the other a proposed sub-city called Aerotechnopolis. This circular area already has a number of manufacturing parks and would contain the airport, new park, and adjacent neighborhoods of Monterrey as well.

Primary Objectives

The primary objectives of the PIIT are to (1) promote applied research and innovation; (2) link research to the needs of the market and companies; (3) use and develop Nuevo Leon's intellectual capital; and (4) incubate new innovation-based businesses.

Main centers in the park include a variety of R&D groups, in operation or under construction. These include universities and public research centers in different fields, including electronics, biotechnology, mathematics, advanced materials, food industry, nanotech, water research, and others. The three major universities, he said, are critical in providing the expertise to compete in these complex new fields. Because of this ongoing research, the park already has an important set of private firms located or planning to locate in the park, including AMD, Motorola, PepsiCo International, Owens Corning, and Infosys. An important ingredient, he said, is partnership with several U.S. universities, including Texas A&M (logistics and mechatronics), University of Texas (IC2, business incubators), and Arizona State University (nanotechnology).

The park is still very new. As of January 2008, the facilities were 20 percent to 100 percent complete. A 150-room Hilton hotel is planned, with facilities for conventions and meetings, a gym, and a swimming pool. Almost 80 percent of the area is said to be committed to prospective tenants. Now under construction are the following:

- CIDESI, Center for Engineering and Industrial Development (mechatronics).
- CIMAV, the Center for Research on Advanced Materials.
- CINVESTAV, the Center for Research and Advanced Studies (medical physics and biotechnology).
- CIATEJ, Research Center for Biotechnology and Food Industry.
- IIE, Research Center for Energy (electrical equipment and renewable energies).
- ITESM-CiDEP, the Center for Research and Strategic Product Design (product design, business incubators).
- Monterrey IT Cluster (42 local software companies).
- Water Institute of Nuevo Leon (recycling methods, technology commercialization).
- UANL-CiiDIT, Center for Innovation, Engineering and Technology Design (advanced materials and nanotechnology, mechatronics, IT and software).
- UdeM, Center of Innovation for Packaging Design and ID Technologies (design services, R&D).
- Also the most important Industrial Groups are committed to install their Technology Centers in the park (CEMEX, ALFA-Sigma, VITRO, METALSA, and others).

Planners envision completion of construction in about two years.

The Park's Main Features

Main features of the park include a total area of 175 acres, investment in infrastructure of $100 million, and investment in buildings and equipment of $150 million. Projected employment over the next five years is 3,500 researchers and engineers. Two business incubators have been designed, one for nanotechnology and one for biotechnology, at a cost of $20 million. The state's first seed and venture capital fund is being assembled by private partners, the government, and the national bank to a level of $30 million. Six years ago, Mexico began providing tax incentives for those who invest in R&D, absorbing 30 percent of annual R&D expenses.

The main challenge, from the government's point of view, is resources. The park will launch this year a new legal framework supporting R&D and innovation, including a 25-year commitment of financial support. For universities, the challenge is to balance the market-driven innovation strategies with the traditional support for basic research. For the private sector, the challenges are to promote an innovation culture, stimulate human resources formation, and provide the scholarships that bring students and faculty to work in private companies. One program, for example, pays the employment costs of a masters or PhD holder to work in a company for 1.5 years.

Mr. Parada concluded by saying that his team has much to learn about the process of transferring expert knowledge into successful business. "We would like to learn from the experience of others with their parks," he said "—about specialized infrastructure, park operation, and successful incubation models. For that reason I am very happy to be here."

DISCUSSION

Mr. Parada was asked how supportive the multinationals already in Monterrey were in supporting the vision of PIIT. He answered that until very recently, Mexico was known for the maquila style of manufacturing,[8] which is "not very attractive." Now, he said, "we are attracting engineering and R&D centers because we can be successful in cost, quality, and human resources. So we are bringing an opportunity to companies already in Monterrey to move beyond that model: to product development, testing, R&D, and technology centers. "We are giving tax incentives, and some additional money. For example, Motorola will probably have its own facility in one year. In the meantime we provide temporary facilities, the student candidates to be hired for training, and all the necessary support for companies interested in taking the next step in innovation."

SCIENCE AND TECHNOLOGY PARK DEVELOPMENTS IN HUNGARY

Ilona Vass
Hungarian National Office for Research and Technology

Ms. Vass, vice president of the National Office for Research and Technology, began by saying that her country of Hungary is "a newcomer to this arena," having become a market economy only a decade and a half ago. She said that Hungary is still among the "catching-up countries in the European Union, with weaknesses outweighing the strengths."

She noted the presence of a strong science base and a very weak industrial

[8]A *maquila* or *maquiladora* is a factory that imports materials and equipment on a duty-free and tariff-free basis for assembly or manufacturing and then re-exports the assembled product, usually back to the originating country.

base, and a "very big mismatch between the two." For example, she said, businesses spend little on R&D, only 20 percent of Hungarian firms are "innovation-active," and the bulk of STI activities are conducted by large, foreign-owned companies. There is poor cooperation between academia and industry, weak commercialization, and a lack of seed and venture capital. Among the country's strengths are a moderately developed business climate, a good science base, and strong R&D capacities.

The opportunities have become more interesting since joining the European Union (EU) in 2004, she said, with structural funding "pouring" into the country. These opportunities include closer integration of multinational corporations into R&D networks, promotion of networking among diverse stakeholders, and integration into ERA Europe, the Electronic Retailing Association.

Threats to the Economy

She noted also "some very real threats" to the economy. For example, there is little cooperation among sectors, low growth, and a largely unproductive pursuit of Lisbon objectives, with a focus on simply spending money rather than on the purpose of the spending.

The nation also suffers from a "dual" economy, she said. To date, the technology development has been driven mostly by foreign direct investment. MNCs and other foreign-owned firms have moved in, and the difference between them and the domestic companies is wide. The foreign-owned firms are typically high-tech, well financed, and do enough R&D to add high value to their products. The domestic firms are technologically underdeveloped, perform little R&D, and are dominated by small and "micro" companies. "There is no way to build a knowledge economy," she said, "with such a backward-looking private sector."

A positive for Hungary is that the growth rate of its R&D intensity is very high and accelerating.[9] "So we have a new and unique opportunity now to promote and accelerate economic growth by investing in R&D and innovation," she said, "—to change from external investment-driven growth to innovation-driven growth, as we have seen in the Asian countries. The good news is that all the stakeholders in Hungary recognize that innovation is the main driver for competitiveness, growth, and creating a knowledge society."

A National STI Strategy

In response, a national STI strategy has been set up for the period 2007-2013. An important element of this strategy is to expand R&D activities beyond Budapest, where "everything is very much centered." Industrial parks are being set up, along with cooperation centers to build relations between industry

[9]Calculated as GERD as a percentage of GDP. Source: DG Research.

and academia. Incubators, services, and technology platforms are planned to strengthen and increase the small number of science-based start-ups and spin-offs. The government is trying to decide where to focus the country's now-fragmented R&D elements and which specialties to build up. It is also trying to encourage business-to-business cooperation, "which is not a tradition." More broadly, the strategy will attempt to better connect science to commercialization, move domestic firms up the value chain, and build critical mass "to be sure we are visible in the world."

Hungary is betting on science and technology parks largely on the basis of experiences elsewhere. As Ms. Vass puts it, the S&T park is a "proven tool to create successful new companies, sustain them, attract new ones—especially in the STI sector—and make existing companies more successful through the use of R&D."

However, the country is under no illusion that this complex development tool is easy to use. She quoted Winston Churchill: "Success is the ability to go from one failure to another with no loss of enthusiasm," noting that for a new economy like Hungary's, it is important to know more about failures in the S&T park field, "because those are what we would like to avoid."

Challenges for Implementing the Strategy

A central question, she said, is how to implement the national STI strategy quickly and in coordinated fashion. "We know that a park has to be a physical location with shared facilities and infrastructure," she said. "It has to be based on private-public partnerships, with a common vision. Unless everyone is included, it's not going to work. This would mean continuing in the same fragmented way."

Although Hungary does not yet have fully functioning research parks, it has about 179 parks with a business or industrial focus, with 2,989 tenant companies and 171,000 employees. These parks, she said, have above average performance in exports and innovation activities. It is not yet clear whether any of these would be transformed into science and technology parks.

A first step has been to set up "regional knowledge centers" to strengthen universities' research capacities and steer their research toward the needs of industry in the regions. These centers plan to integrate university research, education, and commercialization activities into one coherent strategy that involves the private sector as well. One problem is that the 19 centers designated so far may be too numerous, and too specialized; each has a specialty, such as genomics, e-science, cell biology, info-bionics, neurobiology, and medicine.

Providing Innovation Services

Because universities do not have proven strength in innovation services, Hungary has set up Regional Innovation Agencies to provide services like technology licensing and transfer, IPR management, consultancy, technology brokering, and technical facilities. A voucher system was introduced for SME users of services, and two national tech transfer centers were set up and operated as businesses. The University of Texas at Austin was a major partner in setting up one of them.

There are now about 40 business incubators in the country, with an average of 20 tenants and average tenancy of four years. Tenants are mostly in light manufacturing, not emerging technologies. To establish some more modern models, in 2005 Hungary set up a bio-incubator program with two new incubators. In 2006, it established a pre-seed/seed capital fund, which has accumulated about 5 million Euros, and a business angel network.

The country has emulated the European Technology Platform initiative in trying to bring all stakeholders in certain sectors together to set up a strategic research agenda. These stakeholders will then identify key developments needed to raise the impact of that sector on national development. About 11 such platforms are being set up.

The Essential Step of Integration

The next essential step is to integrate the elements of the strategy without allowing one or two to dominate. So far, the park activities have been regional and national, and there is a need to increase their international presence. Finally, the country must mobilize private equity funding and identify the best fiscal measures to stimulate growth in S&T parks.

Ms. Vass said that she favors an approach designed in France of establishing parks as poles of RTDI (research, technology, development, and innovation) activities, including a university research center or research university. Seven such clusters are planned and named for their specialties in major cities, including Technopolis (Miskolc), Innopolis (Budapest), Biopolis (Szeged), Autopolis (Györ), Pharmopolis (Debrecen), Quality of life (Pécs), and Ecopolis (Veszprém-Székesfehérvár). They are based on the regional knowledge centers —universities that have already been strengthened with the goal of better meeting industrial research needs.

A premise of this strategy, she said, is that science parks need to be run as businesses. "That means that in Hungary they will not be run by universities," she said. "The good news is that we have the money available, which we did not have before." The money from EU structural funds will be used to build infrastructure according to the New Hungary Development Plan (2007-2013). Complementary funding, which is necessary for ongoing R&D, will come from the national Research and Technology Innovation Fund, Ms. Vass's employer.

Potential Pitfalls

"We are aware of potential pitfalls in this program," she said, "especially in expecting too much too quickly and over-estimating the park's role because of the high prestige attached to it by the regional governments. Everybody now wants one for themselves, everywhere." A danger, she said, is that the regional governments want to play a major role, but they have little management experience or vision, although some of these qualities will be contributed by Hungarians who have gained management experience in the United States and elsewhere. The parks have already encountered the problem of "too much interference from sponsors," and this she called very difficult, "because the government is providing all the money." She said it is also difficult with multi-channel funding to ensure that needed funding comes to the right place at the right time.

She concluded that Hungary's greatest need in S&T development is to have a clear, coherent, and consistent strategy and, more important, an appropriate environment to set it up and implement it. Old structures are difficult to change, she said, quoting John Kenneth Galbraith: "The conventional view serves to protect us from the painful job of thinking." "The bottom line," she said, "is that Hungary is experimenting with a model that is highly centralized in strategy formulation but decentralized in implementation."

INITIATIVES IN FRANCE

David Holden
Minatec

Dr. Holden spoke about the evolution and current status of Minatec, the first European campus for micro and nanotechnologies, located in Grenoble, France.[10] Minatec is relatively new, an extension of the national laboratory system that has been redesigned to stimulate economic development. The exercise has been sufficiently successful that the Grenoble region is now called the French Silicon Valley, focusing on new products and miniaturized solutions for industry.

Grenoble, said Dr. Holden, in comparison to many places described at the symposium, is a relatively small community of 450,000, and the example of Minatec illustrates what can be done on a local and regional scale. Grenoble was known primarily as a mountain skiing village until just after World War II, when

[10]The full name is Maison des Micro et Nano Technologies.

France started to decentralize some of its research. In 1957, the government selected the site as one of ten national centers dedicated to nuclear research.

Arrival of the Private Sector

A decade later, Grenoble began to change when the Electronics Systems Group, Leti, was formed and gave presence to the private sector. In 1973, Thomson Semiconductor was created as a spin-off, and by 1985 Thomson was number 15 in the world in semiconductor sales. It merged with the Italian company SGS, and in 1988 SGS-Thomson signed a major R&D collaboration agreement with Leti. Soitec, a start-up created by Leti, began business in 1992.

Around the year 2000, as the potential for industry-government high-tech partnerships became clear, and traditional nuclear engineering stagnated, the idea for Minatec was born. It offered a place and a rationale for formalizing new public-private partnerships within a major S&T park that could expand into new fields.

Soon after that decision, Motorola and Philips joined ST Microelectronics in a major R&D alliance valued at 3.2 billion Euros (approximately US$5 billion)—a large industrial investment for a small city. With ST then ranked about fourth world-wide in microelectronics sales, this alliance suddenly became the focal point for Minatec. In 2006, the Minatec campus became the first European campus for education, research, and industry, focused on micro- and nanotechnologies. It now holds ten companies; three are American, and 42 percent of total jobs are related to foreign companies.

A Campus Both Inside and Outside the "Fence"

A major strategic decision for Minatec was to create a campus that is not entirely "inside the fence." University campuses were planned, but, as discussed earlier by Dr. Mote of the University of Maryland, it is seldom easy to do classified or industrial work on a university campus. So a balance has been created between the space allocated for classified work ("inside the fence") and the unclassified activities elsewhere.

Funds were raised for the new park from local and regional governments, and these funds were joined by private and federal investments. To encourage interactions, Minatec made the decision to seek out existing groups and bring them together on the same site, creating a critical mass. Typically, in France, there are many good researchers, but they have been scattered in small groups.

The population of Minatec will soon reach 4,000, including 1,100 students and 1,900 researchers. Of the annual research budget of 320 million Euros, two-thirds comes from outside contracts. The strategic focus is on industrially driven research, based on Minatec's competencies and infrastructure. The 200 industrial partnerships, using 20 joint laboratories on site, produce about

200 new patents per year and 1,360 scientific publications. The target for patents is one per year for every six applied researchers and for publications one per year per researcher.

Facilities are state-of-the-art. Minatec's 300mm silicon wafer center runs 24/7, operated by a public-private partnership. A MEMS 200mm prototyping line allows fast development of new products. Its nanocharacterization platform is unique in Europe, with both in-line and off-line abilities, a very high level of expertise, proximity to device production lines, and synergies between upstream and applied research times. Farther out on the frontier is a dedicated platform for cutting-edge, upstream research. There are major facilities for training and event facilities to bring people together, including a 400-seat conference room, modular meeting rooms, and quick access by foot from the railway station. Minatec hosted some 16,000 visitors in 2007, and a showroom for demonstrating Minatec capabilities to potential partners opened in 2007. "Companies don't really understand Minatec," said Dr. Holden, "unless you show them something."

Four universities on the campus now have 60,000 students, half of them studying the sciences. Three masters programs are offered—in collaboration with other institutions—in nanotechnology (the first in Europe), communications systems engineering, and advanced materials engineering.

Attracting Venture Capital

Minatec has industrial partnerships with companies world-wide, including Bic, Mitsubishi, Total, and Phillips. LETI has a research agreement with Caltech. Contract negotiation, supported by a highly specialized team of engineers and legal experts, has produced more than 250 contracts.

Minatec is beginning to attract venture capital for its start-ups and spin-offs. First-round financing is now being generated in the 15-million-Euros (or US$23 million) range. For example, Crocus Technology, which produces second-generation MRAM products and technologies for semiconductor and electronic systems, received 13.5 million Euros (US$20 million) in first-round investment in May 2006. A spin-off from Leti called Movea, which designs microsystems to capture and quantify human motion, received a first-round investment of 7.3 million Euros (US$11.2 million), and was acquired by Gyration, Inc. in January 2008.

Growing Impact on the Local Economy

Minatec has a growing impact on local economic development. ST Microelectronics has created about 6,000 jobs in the region and is the largest employer. The start-up Soitec has created about 900 jobs, and the other 30 or so start-ups have created thousands more. Altogether, the park produces about 15,000 direct jobs and 30,000 indirect jobs. "We're small enough," said Dr. Holden, "so you can actually do the counting, and be fairly accurate." There has been a net increase of 8,000 high-tech jobs since 2000, which are "the kinds of figures the local authorities like to see."

The Minatec model has also influenced how other S&T clusters in France are financed. The country is trying to decentralize its financial system, with the regions doing more of their own economic development. Competitive clusters are creating groups to screen project applications and then ask for federal money, a process put in place by President Sarkozy several years ago when he was Minister of the Interior. According to Dr. Holden, he used the process at Minatec as the basis for the current French model. Under this system, Minatec received in two years about 1.2 billion Euros (US$1.86 billion) for 113 projects and 315 million Euros (US$487 million) in financing from the federal government.

A New Challenge

In 2007, the Minatec landscape was altered when NXP, formerly Philips, announced it would leave the alliance, and Freescale did the same. ST decided to join the IBM semiconductor consortium developed in Albany, New York, citing the prohibitive cost of microelectronics research. "When certain industries mature," said Dr. Holden, "you have attrition of certain players and much more concentration of effort." France decided at the federal level to purchase 260 million Euros' worth (US$402 million) of ST from the Italian partner to balance ownership and a new 1.2-billion-Euro agreement for ST was signed for 2008-2012 that included IBM. Although the amount of this agreement is smaller, it still represents significant activity. "I think this agreement is good for Grenoble," said Dr. Holden, "because it's focused on technology that is much more diverse and that is better for start-up creation."

A Response: Diversification

The next stage for Minatec, which will be supported by annual inflow of about 320 million Euros, is to diversify into a research triad of micro-nanotech, biotech, and clean-tech (new technologies for energy). The park has added the element of urban development to its strategy, with the approval of the federal government. This element will require an additional 100 million Euros (US$155 million) from the government, which is currently contemplating an even more ambitious model for the park and its relationship to the city. "It is not a very conventional S&T park," concluded Dr. Holden. "It is very active in the development of Grenoble and all of its urban development programs, so this new project meshes with what will happen in transportation and new tram lines and expressways, as well as science. So it's very easy to sell when you have to go and ask for money."

DISCUSSION

A questioner asked whether all the Minatec investment goes into pure R&D, and how jobs were created. Dr. Holden replied that in order to participate in Minatec, a partner has to have a research contract with them, but that job creation happens in many places. Some of the start-ups are in California; some have offices in both Grenoble and California. "There are no restrictions on that. In many ways, I don't think that putting requirements on jobs is the best way to be successful. We find that typically the start-ups located in California still do R&D with us. So for the local economy, which is research focused, it's beneficial."

Another questioner asked about comparisons with the Hungarian experience. He said that in both France and Hungary, as in the rest of Europe, it is typical for the government to help with the infrastructure. This is traditionally considered a public responsibility, meant to encourage the private sector to move in.

Dr. Parada noted that in Mexico as well, the government leadership and financial support is critical. Once the government creates the necessary infrastructure, the park can attract private investors who put money into the research parks and innovation ventures.

Dr. Holden added that in Grenoble, of the 3.2-billion-Euro investment, the local government put in about 150 million Euros, most of it to pay for infrastructure, such as highways and access roads. This investment, he said, has been more than paid back in the form of corporate taxes over the four-year period, and the local government is still benefiting from a net positive of 1,000 technical jobs and perhaps three times as many support jobs.

A questioner agreed that local job creation is desirable, though not the best true measure of park success, and asked what other metrics Minatec used. Dr. Holden said that the GDP of the tech sector is fairly easy to track, highlighting the impact on company business, growth rates, and salary levels. He also said that education is a good metric: "Here we see that we've tripled the number of engineers in the past decade, so that's pretty significant. In the knowledge economy you have to have an educated population, and even though we may be draining other parts of France and Europe in some ways, we are creating this critical mass, and, to be competitive, that's what's necessary."

Dr. Wessner noted that the leading centers for silicon research today are no longer confined to Silicon Valley; they are in Austin, New York, Flanders, and Grenoble, and they all used public money to begin. He suggested that those local governments clearly recognized the value of the investments they were making. In Grenoble, for example, when conditions changed, the government increased its investments rather than shutting them off, and the park shifted its strategic planning to take the new conditions into account.

Keynote Address III

Introduction

Charles Wessner
National Research Council

Introducing Dr. Barker's keynote address, Dr. Wessner noted that the role of research in universities is evolving. While the traditional activity of academic scientists is to perform basic research, recent evidence suggests that academic scientists who collaborate with industry excel in both spheres. This suggests, he said, that universities can play a valuable role in innovation and in the commercialization of knowledge than previously assumed.[11] President Barker's achievements at Clemson University, he said, would provide additional evidence for this point of view; Clemson has raised both its academic standing and its involvement in economic development during his tenure.

James Barker
Clemson University

President Barker began by praising the work of the National Academies and the Association of University Research Parks and "the leadership of both groups in creating opportunities for policymakers, academic leaders, and scientists to engage with each other in a common purpose." The common purpose of the two institutions, he said, is that science and technology can advance knowledge "and stimulate economic development, create jobs, and improve the quality of life for

[11]Bart Van Looy, Marina Ranga, Julie Callaert, Koenraad Debackere, and Edwin Zimmermann, "Combining Entrepreneurial and Scientific Performance in Academia: Towards a Compounded and Reciprocal Matthew-Effect?" *Research Policy* 33(3):425-441, April 2004. The authors found that groups that collaborate have a reinforcing effect and generate more fundamental scientific output as well as developmental research, as measured in number of publications.

all the world's citizens." During this time of economic uncertainty, he said, both institutions recognize that their work together is more critical than ever. Concrete solutions to such global issues as renewable energy, safe food supplies, climate change, disease prevention and treatment, and even security are likely to be found on campuses through science and technology advancement, and in many cases in S&T parks.

THE ADVANTAGE OF COLLABORATION

There will always be "eureka moments" from solitary researchers working alone in a lab, he said, but we are now discovering that collaboration can be a tremendous competitive advantage. He said that Clemson has worked hard to make collaboration a part of campus culture. In this climate, the Clemson University-International Center for Automotive Research (CU-ICAR) has been established on a 250-acre campus to create opportunities for collaboration. Designers have intentionally created a "high-density campus" so that researchers from university and industry could not help running into each other: They share parking structures, workout facilities, and eating establishments. "It is a physical manifestation of a core Clemson belief—which is that innovation is a contact sport." A result is that the intersection of university and industry accelerates the transfer of research to the marketplace.

President Barker gave some of the context for Clemson's collaborative culture. The university was founded in the 19th century and named after a man from Philadelphia named Thomas Green Clemson—an engineer, scientist, artist, and musician educated in Paris. He served as U.S. Secretary of Agriculture, and, in Washington, he met and fell in love with the daughter of statesman John C. Calhoun of South Carolina. He returned with her to her home, which is today a part of the Clemson University campus. He worked to establish a university that would restore the weakened Civil War economy by providing advanced education and scientific experimentation in agriculture and engineering. At his death he left his land, home, and personal fortune to support the founding of Clemson University, writing in his will: "I trust that I do not exaggerate the importance of such an institution for developing the material resources of the state." The university, then, was founded specifically as a driver of economic development.

A UNIVERSITY COMMITMENT TO ECONOMIC DEVELOPMENT

For 120 years Clemson has followed that vision in first serving the state's key industries of agriculture, textiles, and ceramics, and more recently sectors such as automotive, advanced materials, and biotechnology. In the 21st century, CU-ICAR is a physical manifestation of Clemson's commitment to economic development through innovation. Four years ago, CU-ICAR was 250 acres of undeveloped land along the I-85 corridor between Charlotte and Atlanta, with no master plan, no business plan, no curriculum, and no funding ("a typical university project").

Today CU-ICAR includes a 90,000-square-foot graduate engineering center with world-class faculty holding well-funded endowed chairs. It is the only location offering masters and doctoral degrees in automotive engineering. CU-ICAR houses R&D facilities occupied by BMW and Timken, and has active new partnerships with Michelin, IBM, Dale Earnhardt, Inc., Sun Microsystems, the Society of Automotive Engineers, and the Richard Petty Driving Experience. It is also the first North American home of two international software companies with major automotive and motorsports clients. In all, CU-ICAR has generated more than $220 million in public and private investment and has created more than 500 new jobs with an average salary of $72,000. "All of this in only four years," he said.

AN AUTOMOTIVE PLATFORM FOR INNOVATION

A critical ingredient in the success of CU-ICAR has been the decision to focus on the automotive sector as a platform for innovation. President Barker described three reasons for this decision:

- Vehicle-related R&D has been an area of academic strength since the 1970s when faculty were gaining a reputation in vehicle dynamics and control systems applied to rail vehicles. Model developing, testing, and other research interests evolved into automotive engineering in the 1980s, and into motorsports engineering in the 1990s, where Clemson has been a pioneer. In areas related to vehicle systems and dynamics, Clemson's faculties have been national leaders.
- Second, the automotive and motorsports sector has been identified by the state as critical to the state's current and future economic prosperity. In the 10-county region around Clemson there are 125 automotive suppliers and related companies, and statewide there are six original equipment manufacturers (OEMs) and more than 1,800 automotive-related manufacturing plants and companies. There are more OEMs and first-tier suppliers within a 500-mile radius of Clemson than in the same radius around Detroit. Further, the Upstate as a region is located directly along the U.S. motorsports racing corridor. Two-thirds of U.S. racing teams are located between Atlanta and Charlotte, and Clemson is exactly mid-way between those two cities. The Charlotte-to-Atlanta I-85 corridor is the eighth largest regional economy in the world,[12] and enhancing this cluster is as much a part of Clemson's core mission as was rebuilding the war-ravaged state economy in the 19th century.
- Clemson has a strong industry partner in BMW "which believes that

[12]Richard Florida, Tim Gulden, and Charlotta Mellander, "The Rise of the Mega-Region," October 2007.

innovation is the key to future financial stability and is willing to make a significant long-term investment to secure that future." Their investment, in endowing chairs and in economic development incentive funds directed at CU-ICAR, "was not an act of philanthropy," he said, "but a conscious business decision by a company." Clemson faculty work side-by-side with counterparts from BMW and other firms in developing "curricula relevant to current and future industry needs, building in such things as language and international study experiences and incorporating industry experience in emission requirements." President Barker recalled news media reports that have questioned the close relationship between Clemson and BMW, but said there have been no efforts to influence faculty hiring, curriculum, or admissions decisions.

AN EFFORT TO INTEGRATE DISCIPLINES

President Barker differentiated between "being focused and being narrowly defined." He said that CU-ICAR focuses on the automotive sector but is not limited to manufacturing. The concept of the program is that "the automobile can serve as a platform for innovations that can be translated to countless other products and manufacturing processes. Unless you're a commercial airline pilot, the automobile is the most complex system you'll ever operate. It includes hundreds of electronic and mechanical components that have to interact constantly and flawlessly with each other and with the external environment, the surface of the road, and human occupants." The broader goal of the education and research program is to create a workforce of systems engineers—"integration engineers, if you would." Traditionally, engineers have worked deep in their "one-disciplinary holes," he said, and the challenge has been to "tie those holes together—to produce people who understand and improve how extremely complex systems interact with each other and apply these principles to a broad spectrum of applications."

THE ESSENTIAL ROLE OF THE STATE

He acknowledged that even with those three factors—a strong faculty, a clearly defined need, and a powerful industry partner—Clemson would not have been able to establish CU-ICAR without state funding. For many years, he said, South Carolina "watched from the sidelines" as neighboring states moved ahead in per capita income, job creation, and development of high-tech industry by leveraging the strength of their state universities. In 2002, the General Assembly decided it was time to catch up and passed a series of legislative programs to encourage Clemson and the other two research universities to accelerate their economic efforts.

The timing was right. BMW was also expanding, and when it increased its capacity by investing $400 million and adding 400 new jobs, it qualified for more than $100 million in state incentive funds. The company directed $25 million of those funds toward construction of the graduate engineering center of CU-ICAR and another $15 million toward building a new IT facility adjacent to CU-ICAR. Also in 2002, the state Research Centers of Economic Excellence Act set aside $30 million annually in lottery revenue to fund endowed chairs in areas related to economic development at the three research universities—provided that the university generated matching funds. With private gifts from BMW, Michelin, Timken, and other partners, this program has invested $36 million in endowments to support faculty and students in graduate engineering centers of CU-ICAR. In 2004 the Research Universities' Infrastructure Act offered $210 million for facilities and equipment that these faculty and students would need, once again requiring matching dollars from non-state sources. Clemson received more than $38 million in state funds which would be matched by private sources to help build CU-ICAR's physical plant and infrastructure.

These core pieces of legislation, President Barker concluded, created "instant scale and instant density." They gave CU-ICAR momentum that might otherwise have taken decades to develop. They also provided a high level of accountability through peer review and oversight by a board of industry and economic development leaders who made many of the decisions about how to spend those funds. "I can't overstate the importance of effective public policy and designated public funding," he concluded. "Otherwise we would have had a great idea but no means of turning this idea into reality."

DISCUSSION

A questioner asked how Clemson would continue to build on its fast start. President Barker said that the university has a master plan of five neighborhoods and was just completing the first of those five. The next challenge would be to continue that momentum in the face of the current economic downturn. Another challenge, he said, is the wide difference between the pace of government and the pace of private companies. "We're in the middle trying to sort through that, another big challenge." He said he had imagined CU-ICAR as existing "in the overlap between the academy and corporations. The more I looked at the diagram, the less overlap I saw. There is no overlap. The challenge is building a bridge between the two circles."

He was asked about the importance of his own training as an architect. He said that it was critically important, because his education was in "placemaking," which he sees as a vital part of the success of a park. "We wanted to make sure that each building won silver LEED certification and an AIA design award," he said. "It has let us see that we have to work harder on our housing component. We have eating establishments, workout facilities, parking structures, but you really don't have placemaking without 24-hour occupancy."

A participant questioned whether the automobile is an appropriate platform to build on in times of energy and environmental concerns. President Barker said that it was, because automotive research intersects all current challenges—energy consumption, pollution, green issues, hybrid technology, and safety. "As solutions to those issues are found—and I hope they are found on our campus—they will influence the kind of research being done in many contexts." He added that CU-ICAR faculty "are not working on research that's part of the past, but part of the era just beginning."

A questioner asked how Clemson's state has managed to move so quickly. "There's nothing magic about South Carolina," answered President Barker. "But one lesson for me was during the last economic downturn the university's academic budget was cut by about $50 million, and at the same time, the state invested $50 million in economic development funds that came directly to Clemson." He saw this as a "net shift from investing in basic education to investing in a knowledge-based economy that reflects the priorities of the state. What's happening to universities around the world is that we still have a responsibility to support the intellectual development of our students—to study Aristotle and ask questions about beauty and truth and the meaning of life. But we also have to do economic development with the state. The trick is to figure out how to do both simultaneously, and be wise about funding. It's like driving a car on the interstate at 80 mph while you redesign and rebuild the car."

He was asked about the role of the federal government in making investments in innovation. He said that Clemson has received only a small amount of funding from Washington. "If we're going to be competitive world-wide," he said, "we as a nation have to build a knowledge-based economy. How best to help that at the federal level may be to look at successful models at the state level and decide which of those should be supported."

U.S. Parks: The Laboratory Model

Moderator:
Kathryn Clay
U.S. Senate Energy Committee

Ms. Clay said that the Senate Energy Committee has a dual mission in trying both to support work on basic science and also "to see the importance of following through on basic research to technology transfer and commercialization of products."

She noted that Committee members Bingaman and Alexander have worked hard on the American Competitiveness Act, which is based on recommendations made by the National Academies' Augustine Commission. This report, Rising Above the Gathering Storm, laid out a rationale for federal support of basic science and science education.[13]

In addition to basic research, she said, the Senate Energy Committee has long advocated that the nation's national laboratories do more than produce knowledge—that they also help turn new knowledge into products for the marketplace. This broad mission of technology transfer, she said, includes taking projects that begin on the bench and helping them "come off the shelf" to become useful technologies. This remains a difficult balance for Congress because of the wide range of views about whether a federal role in commercializing technology is appropriate. "While almost any member of the Senate will be supportive of the

[13]National Academy of Sciences/National Academy of Engineering/Institute of Medicine, *Rising Above the Gathering Storm: Energizing and Employing America for a Bright Economic Future,* Washington, DC: The National Academies Press, 2007.

government role in basic research," she said, "it is often difficult for people to understand the subtle difficulties of getting that research to the marketplace."

Another way the federal government can help propel research results into the marketplace is by investing in science parks. She said that Senator Bingaman has been a champion of legislation for science parks, and that Sandia National Laboratories has become a model federal effort in promoting the public-private relationships of successful parks.

U.S. AND GLOBAL BEST PRACTICES: SANDIA SCIENCE AND TECHNOLOGY PARK

Richard Stulen
Sandia National Laboratories

Dr. Stulen, chief technology officer at Sandia, began by noting that Sandia Science and Technology Park has many features in common with Minatec in France, including deep roots in nuclear engineering research. Sandia National Laboratories were established in New Mexico in the late 1940s, under President Truman, to develop nuclear weapons. While Sandia remains a national security laboratory, its mission has broadened into other national security arenas, including energy and microelectronics, which rest on a broad base of science, technology, and engineering research. Dr. Stulen said that it is this juxtaposition of science and engineering that distinguishes Sandia from other national labs in the country.

Sandia: An Attractor for Industry

The new Sandia Science and Technology Park has grown out of that research base and sits at the opposite end of an "innovation corridor" from the laboratories. Between them is the multi-building complex of MESA, Microsystems and Engineering Sciences Applications. This corridor, which represents a $500 million investment by the Department of Energy (DOE), resembles Minatec in that it extends from inside the classified area to a nonclassified region. And like virtually all the parks described in the symposium, said Dr. Stulen, Sandia's access to special people, knowledge, capabilities, and equipment provides a strong attractor for industry.

For example, Sandia has made a large investment in high-performance computing (HPC). Its Red Storm HPC Facility is one of the most powerful in the world. Adjacent to several MESA buildings, it offers valuable capacity for modeling and simulation. MESA also has a microelectronics fabrication facility and capabilities to do mixed-mode fabrication using both silicon, III-V compounds such as GaAs, and microsystems, said Dr. Stulen, providing the ability to integrate single chips in many ways not normally available to industry. The Joint Computation and Engineering Lab brings modern simulation capabilities to the design process in a seamless way that accelerates discovery design for Sandia and for some of its industry partners, "which is a critical ingredient into the future." Farther out from the National Laboratories is a new Center for Integrated Nanotechnologies, a partnership with Los Alamos National Laboratory, which is one of five nanoscience centers funded by the Department of Energy. It is outside the "fence," available to research partners from around the world.

Several more Sandia facilities are located in the research park itself, which is separated from MESA and Kirtland Air Force Base. Part of the Computational Science Research Institute has recently been placed in the park, and other activities are being integrated as more of the laboratory moves outside the fence to form partnerships with universities and companies moving into the park.

Unusual Partnerships with the Landowners

The 240-acre park was founded in 1998 to attract industry in support of the Sandia mission. The park is unusual in having three founding partners: Sandia, Technology Ventures Corporation, and the city of Albuquerque. Their common vision has been to attract business that would co-locate, to the extent possible, with the laboratory. Another unusual feature is that the park does not own the land it occupies, so an MOU had to be signed with the landowners—the Albuquerque Public Schools, BUILD New Mexico/Union Development Corporation (private), and the New Mexico State Land Office.

The initial purpose of the park was to create joint research and development opportunities, commercialize technologies, bring in new business, strengthen supplier-based "collaboratories," and foster regional economic development. "These are more or less the common objectives for all parks world-wide," said Dr. Stulen. While its goals were similar to other parks, however, Sandia Park does have a competitive advantage in its proximity to world-level expertise and infrastructure.

The Need for Champions

Dr. Stulen emphasized, however, that expertise and infrastructure are not sufficient to ensure success. It also requires one or more high-level champions, people who care and have the ability to direct resources continuously to the park. "Parks don't just happen," he said. "They require energy, devotion, passion from leaders—not only of the institution but also of the region." He praised not only Senators Domenici and Bingaman for that leadership, but also political leaders at the state and local levels.

He emphasized also the importance of many kinds of public-private partnerships. In addition to the landowners, for example, other key partners were the

Park Metrics		Results Since Park Founded in 1998
Number of Companies		27
Number of Employees		2,113
Number of Buildings		18
Square Feet of Occupied Space		897,925
Acreage Developed (out of 240)		67
Funds-In and In-Kind Services from Tenants to Sandia (i.e. CRADAs, Licensing Agreements)		$17,591,682
DOE/Sandia In-Kind Services to Tenants CRADAs)		$2,667,916
Contracts from Sandia Procurement to Tenants		$244,468,938
Contracts between Tenants		$7,186,865
Public and Private Investment in the Park	Public	$66,811,090
	Private	$229,369,458
	Total	$296,180,548
Average Salary for Each Full-Time Job in the Park		$62,000
Average Salary for Each Full-Time Job in Albuquerque		$37,000

FIGURE 3 Focus on results.

Department of Energy, Lockheed Martin Corp., Economic Development Administration, State of New Mexico, Bernalillo County, Public Service Company of New Mexico, Mid-Region Council of Governments, and the New Mexico congressional delegation. "We've had help from a range of levels, at every scale."

He offered some statistics to give the dimensions of the park: 27 companies, 2,113 employees, 18 buildings, 897,000 square feet of occupied space, 67 developed acres. Funds-in and in-kind services flowing from tenants to Sandia, such as CRADAs and licensing agreements, have totaled $17.6 million, and DoE/Sandia in-kind services to tenants (CRADAs) have totaled $2.7 million. In the other direction, contracts from Sandia procurement to tenants amount to $244.5 million. "The funding goes both ways," said Dr. Stulen. "This demonstrates what partnerships can do."

Successful Tenants

As an example of successful tenants, the first company founded in the park was EMCORE, in 1998, which is publicly traded. It has licensed three kinds of technology from Sandia: VCSEL, vertical cavity surface emitting laser technology used in telecommunications; solar cell technology; and transponder technology. It has invested $103.8 million in the park and has 500 employees.

Another successful park tenant is KTech, a local company that provides technicians for the Sandia Pulsed Power Facility. KTech has invested $34.2 million in the park and also has 500 employees.

The park has received $5 million in grants—"small compared with some of the Asian parks, but we consider it significant," he said. Just over half came from the U.S. Department of Commerce's Economic Development Administration for a fiber optic backbone and equipment; over $1 million came from the state for infrastructure; and $785,000 came from the city of Albuquerque.

Dr. Stulen closed by summarizing some challenges now faced by the park. Foremost is the "continuing challenge of keeping the federal government engaged, and maintaining their interest in providing incentives to help small businesses work in the parks." He also said the park needs to improve mechanisms for industries to engage with the national laboratories. "We need more speed in working with industries," he said, "to be able to work at their pace. This means cutting through the bureaucracy in doing an IP agreement, meeting governmental regulations, and measures. We still have challenges there."

NASA RESEARCH PARK

Simon (Pete) Worden
NASA Ames Research Center

NASA Ames Research Center in California—one of ten NASA centers—has an unusual history. It began as a 500-acre NASA property in 1939, to which an additional 1,500 acres was transferred following the deactivation of the Naval Air Station at Moffett Field in 1994. Since 1998, NASA has sought to develop the NASA Research Park (NRP) on the property, with the goal of creating a world-class, shared-use S&T campus for government, academia, nonprofits, and industry. NASA Ames has spare capacity; still undeveloped are several million square feet of old Navy facilities, including the three huge dirigible hangers that once held the ill-fated USS Macon and other airships of the 1930s.[14]

The NRP is well situated to take advantage of high-tech partners from all sectors. Adjacent to Sunnyvale and Mountain View, California, some of the nation's leading technology companies are neighbors, including Google, Hewlett-Packard, Apple, and Intel. The park's mission is to support NASA's mission through research partnerships with universities, industry, and other research groups.

[14]When the airship USS Akron crashed off New Jersey in 1935, one victim was Admiral Moffett, after whom Moffett Field was named. The admiral was an ardent proponent of the airship model, and Hangar 1 at Moffett Field was built to house the USS Macon, the sister ship of the Akron. The hangar is a fifth of a mile long, and so high that fog sometimes forms in its upper reaches.

Successful Collaborations and Spinoffs

"We think this has been a real success story," said Dr. Worden. "We already have more than 40 industry and 14 university partners onsite. We have several million square feet of old Navy facilities, including the dirigible hangers. We have many successful R&D collaborations and spin-offs, support from our congressional delegation and the U.S. government, a NASA-approved business plan, and an approved environmental plan."

The NASA Research Park, he said, has three key objectives:

1. Support NASA's mission.
2. Generate revenue for the Ames center, using an enhanced use leasing program to convert underutilized lands.
3. Raise political and public support for NASA by providing benefits to the local economy, public education, and the U.S. scientific base. "We want to boost NASA's prominence before Congress and the public, strengthen community ties and investments, and broaden NASA's relationships with industry."

In 1997, the towns of Mountain View and Sunnyvale appointed a Community Advisory Committee to come up with appropriate uses for Moffett Field. They recommended such measures as air shows, information technology institutes, astrobiology institutes, an S&T campus and light industrial park, an air and space educational center, expansion of the Bay Trail, film studios, and expansion of the U.S. Space Camp. All such projects are likely to advance NRP's third objective of building visibility and serving the community.

A series of actions during the years 1998-2002 set the tone and framework for NRP as it is today:

- In 1998 a planning MOU with Sunnyvale and Mountain View was signed to establish the research park.
- In 1998-1999 MOUs were signed with the University of California (UC), Carnegie Mellon University, San Jose State University, and the Foothills-DeAnza Community College District. It also has a partnership with the United Negro College Fund.
- In 2000, NRP began developing an Environmental Impact Statement (EIS).
- In 2002 a final EIS was approved, allowing for 4.2 million square feet of new construction. The EIS received many compliments for its quality. NRP has also received several awards and positive reviews for its development planning, partnerships with industry and academia, and probable return on investments.

Development Plans

The Space Act of 1958 gives NASA power to sign agreements that are not covered by any other law or regulation. This power has been used extensively in developing the campus. NRP's development approach has also won praise, and Ames has been selected competitively within NASA for enhanced use leasing. This form of leasing increases flexibility, retains rent payments at the center, and can accept in-kind considerations.

Its research programs, in addition to the educational institutions mentioned above, include CREST (Center for Robotic Exploration and Space Technology), M2MI (machine to machine intelligence), Bloom Energy (fuel cells), and UAV (unmanned autonomous vehicles.) It also includes collaborations and contracts with many small firms. In 2005 the park signed an MOU with Google to build 1 million square feet of new facilities for large-scale data management and collaborations in massively distributed computing and bio-info-nano convergence. In 2006 a NASA-Google Space Act Agreement for Research and Development Collaboration was signed, with plans for up to 100 rental units of housing on 40 acres and new R&D labs.

NRP has also begun discussions to build a major campus with a consortium of universities, led by UC Santa Cruz. The consortium will lease about 70 acres for research, education, and innovation. Goals are to develop new technologies emerging from the convergence of bio-info-nano-scientific research, autonomous systems and robotics, renewable energy sources, technologies for long-term sustainability of human life, and managing innovation in the emerging world. The project," said Dr. Worden, "may evolve into the 11th campus of the University of California."

"These emerging programs are quite exciting," he concluded—"a work in progress. We feel that we've done a lot to enhance the visibility of the park in the private sector and the entrepreneurial community, and this in turn has helped us a lot in moving toward our research goals."

THE NATIONAL CANCER INSTITUTE AND NCI-FREDERICK

John Niederhuber
National Cancer Institute

Dr. Niederhuber, director of the National Cancer Institute (NCI), began by saying that NCI is one of the National Institutes of Health (NIH), founded in 1937 as the first disease-based Institute. He said he would discuss the NCI plan to bring together much of its technology research and development in a park-like setting in Frederick, Maryland, "which I believe will enhance opportunities to interact with the private sector."

The public wants NCI to do three fundamental jobs with respect to cancer, he said: (1) diagnose the disease before it becomes a disease, (2) develop new ways to prevent it, and (3) develop new therapeutics that would "change it from a death sentence to a chronic illness with good quality of life." To achieve these goals will require "very basic science"; identification of targets and pathways within cells; interventions using small molecules, large molecules, and biologics; processes of chemistry; high-throughput library screening; testing therapies in animal models; and eventually clinical studies.

"That I see as the center of what we do," he said, "so I think you can understand how that requires a platform involving the academic environment, the private sector, and the federally funded NCI. This really is a platform of connectivity that enables all the centers we work with."

Accelerated Pace of Biomedical Discovery

He said that in the biomedical sciences, the pace of discovery and generation of new knowledge is more rapid than it has ever been, in part because the Human Genome Project precipitated a revolution in the ability to study biological systems. One scientist to recognize what was to come, he said, was Walter Gilbert, who won the 1980 Nobel Prize in chemistry. Gilbert stated, "The list of genes that will come from the genome project will be the tool that turns our questions into global ones."

This tool is rapidly being applied in genome-wide association studies, where large populations with disease are genetically compared with those of the population who have not contracted disease. The goal is to find single nucleotide polymorphisms that can be used to show, for example, why one person differs from another in the risk for developing certain kinds of cancer or other conditions. NCI has already completed extensive association studies on breast, prostate, lung, and colon and now is close to completing work on pancreatic cancer.

The Institute has also begun to extend its genetics work to sequencing whole tumors and developing a human genome of the cancer. Pilot projects are underway in lung cancer, ovarian cancer, and glioblastoma to help develop the technology needed to do such sequencing and to learn whether it can be as beneficial as expected. He said that the preliminary work on glioblastoma is very exciting, and some of it will be released within the next few months.

Recent studies have also moved beyond a focus on the cancer itself and examining the cancer as it resides in normal tissue. This is because it is now recognized that the "normal" tissue has a role in the disease as well. While it may look normal under the microscope, there's a dynamic relationship between it and the tumor. There is also an interest in whether the tissue stem cells—generated, for example, in the lining of the intestinal tract—are uniquely susceptible to genetic changes that may lead to the cancer process.

The Role of NCI-Frederick

These changes reflect the increasing complexity of cancer research. The word "cancer" really refers to hundreds of diseases: many breast cancers, many colon cancers, all genetically different. Understanding them will require more knowledge about genetic changes and differences than about the organ site.

Because of this research intensity of modern cancer studies, NCI has developed a unique resource in its NCI-Frederick campus to take advantage of more partners and a wider knowledge base. Established in 1971 by President Nixon, using land from Ft. Detrick,[15] NCI-Frederick is, said Dr. Niederhuber, "much more than a satellite campus." It was established as a rapid response site to develop new technologies to support the "War on Cancer" proclaimed by President Nixon. It was given FFRDC[16] status in 1975, which allows NCI to establish contractual relationships in streamlined fashion. Of the 38 FFRDCs, NCI is the only one devoted to biological research. "I believe this is why we have made the progress we have," said Dr. Niederhuber.

Now NCI is developing its presence at Ft. Detrick in the direction of an S&T park. Part of NCI's goal in managing NCI-Frederick is to enhance its partnerships with the private sector and academia and to facilitate various integrated research programs. These objectives, as the symposium heard from many speakers, are brought closer by the rich collaborative environment of a park.

The Potential Benefits of a Park

In particular, said Dr. Niederhuber, a park holds out several potential benefits for NCI:

- Some 82 percent of the NCI budget is spent extramurally to support research at research universities. The Institute has realized that it could make this spending even more effective by creating a science and technology research park close to the Frederick campus.
- The proximity of the campus would allow NCI to improve productivity and reduce the costs of drug development and diagnostics.
- A park might enable execution of larger and more complex experiments and provide a better environment for training the next generation of people to work in cancer-related technology development.
- It can offer incubator and think-tank space to integrate the work of multiple

[15] Ft. Detrick, during and after World War II the center of biological warfare research for the Department of Defense, today supports a large multi-agency research community under the authority of the U.S. Army Medical Command.

[16] FFRDCs, federally funded research and development centers, are not part of the government, but are operated by universities and nonprofit corporations under federal contracts. They address national research needs by bridging the public and private sectors.

companies and research teams with significant R&D programs in engineering, software, and imaging.

To bring this plan into reality, NCI has been negotiating with developers to create a park of about 230,000 square feet, and Dr. Niederhuber expects to be able to make a formal announcement in the summer of 2008. According to the plan, NCI will be the anchor of the park, but it will support the broader mission of technology development for biological research.

Creating a "Dynamic Intellectual Environment"

"This is not simply a way of enhancing our infrastructure," said Dr. Niederhuber. "We want to create a dynamic intellectual environment, hopefully in neighboring facilities of the NIH, other federal laboratories, and the private sector."

He said that he often ends his talks by reminding people that NCI's results from cancer research are not restricted to that disease. "If you think about advances made in almost every area of disease," he said, "you can trace them back to work done using cancer as a model of disease. The research involves finding the most fundamental ways of accessing tissue and using that access in innovative ways. Much of what we know about all disease continues to be derived from work using cancer as a model of biology."

DISCUSSION

A questioner asked whether NCI's FFRDC status helps other parts of NIH, and whether NASA Ames has sufficient authority from the Space Act to act quickly. Dr. Niederhuber agreed that the FFRDC status did help others, and cited the example of a large vaccine facility NIAID was able to build in only 18 months with NCI help. "It helps virtually every institute on the NIH campus," he said.

Dr. Worden said that in the case of Ames, about half the employees were civil servants, but attempts to gain FFRDC status had not succeeded. "One big problem," he said, "is that when one administration wants to change direction, there is a lot of inertia." However, he agreed that the Space Act provides considerable flexibility.

Wendy Schacht of the Congressional Research Service asked about IP issues in dealing with partners and tenants in parks. Dr. Stulen said that IP is sometimes difficult to manage, but remained an important tool in attracting and negotiating with park tenants. Dr. Niederhuber said he envisioned some IP complexities in future drug development. "We are going to need to prove the efficacy and safety not of a single agent at a time, but a recipe of highly targeted therapies that could have multiple IPs, and we're not sure how that's going to work. We may require some policy legislative assistance down the road. Other issues about IP are fairly straightforward as long as you keep them up front."

Another questioner asked whether the national laboratories have found ways to allow investigators to be actively involved in start-ups. Dr. Worden said that this is indeed a difficult issue at Ames. Some scientists left government to work on start-ups and, if they did not succeed, often came back to work at Ames. He said that a legal advisory program was needed.

Dr. Stulen said that Sandia has an entrepreneurial separation program that frees people to leave for up to two years to establish a start-up, with an option to renew for one more year and then return to the lab should things not work out. This helps scientists take the risk necessary to start a new company and provides a way for Sandia to attract high performers back to the lab. Venture capital firms, however, prefer that scientists sever their link with the lab to commit completely to the new venture. Dr. Niederhuber said that NCI did not have "anything close to that, and recruitment and retention are more difficult because of it. I think the pendulum has to come back a little."

Dr. Wessner asked Dr. Niederhuber about the anticipated financial scale of the new NCI-Frederick Park and whether NCI would follow the practice of partnering with a university. Dr. Niederhuber replied that some of the investment took the form of labor by himself and others at NCI. In terms of capital expenses, he planned to have a developer construct the building, which NCI would lease back. Of the total NCI budget of $4.8 billion, some $17 million is used for leasing laboratory and office space, and some of this will be used at Frederick. Infrastructure and equipment will also come out of the existing NCI budget. Private partners will be asked to finance their own facilities.

He said that the partnerships with universities would be "facilitative." That is, an investigator might come forward with a project—perhaps a cellular target that has promise for treating cancer. The project might need some advanced analytical techniques that NCI could offer. For example, high-throughput screening of libraries might be required to see if the compound has promise. Or significant chemistry input might be needed to refine a small or large molecule, optimize it, and determine how it could be used in formulation. NCI could help a university partner do this testing and also study the molecule in terms of how it will change function and whether it has efficacy. "That's the kind of partnership we imagine," he said. "We can also do a great deal to facilitate the interaction of academic investigators with the private sector."

U.S. Parks: University-Based Models

Moderator:
Christina Gabriel
The Heinz Endowments

The Heinz Endowments is a $1.6 billion, Pittsburgh-based philanthropic foundation. The Endowments focuses its philanthropy on southwestern Pennsylvania, but in supporting the regional economy, said Dr. Gabriel, Heinz addresses "issues that are national in scope." A challenge for the Pittsburgh region is to continue its transition from a steel-based economy to an innovation-based economy. Heinz promotes that transition by using philanthropic dollars to stimulate the regional economy. "All has to be done in partnership," said Dr. Gabriel. "We're well along the way, but we're not there yet."

She then turned to invite Rick Weddle to the podium.

RESEARCH TRIANGLE PARK: PAST SUCCESS AND THE GLOBAL CHALLENGE

Rick L. Weddle
Research Triangle Park

Rick Weddle, president and CEO of the Research Triangle Foundation of North Carolina, described the Research Triangle Park (RTP) as centrally located among its three university partners—the University of North Carolina at Chapel

Hill, Duke University, and North Carolina State University. RTP was founded in 1959 by business, government, and academic leaders, he said, although the universities were the key drivers of the concept.[17] They saw the importance of stemming the brain drain from North Carolina and promoting a shift in the regional economy from agriculture, low-wage manufacturing, and government to a research and technology focus.

Employment growth began at the park in the mid-1960s, a time when less than 12 percent of the employment in the region was in high-tech industries. After climbing slowly to around 10,000 jobs in the 1970s, the employee population grew rapidly during the 1980s and has continued to grow since then. Mr. Weddle projects about 45,000 S&T jobs in the park by 2016, spread among some 160 S&T-based firms.

ECONOMIC IMPACT OF RTP

The park has had a long-term economic impact on the region. Per capita income growth in Raleigh-Cary and Durham were far below the state average and national averages before the park was formed; today the per capita income of the region significantly exceeds the U.S. average and far exceeds the North Carolina average. In the 1960s it was one of the poorest regions in the southeastern United States and today is among the wealthiest regions in the southeast.

Mr. Weddle noted that "scale is important" and that RTP is a large park at 7,000 acres. The park did a survey and found that the average size of a research park among the membership of the Association of University Research Parks in the United States is 358 acres, and the average size of a park in the International Association of Science Parks is 708 acres.

However, the average size of a park in China is 10,357 acres. "I thought we were a big park until I looked at that," he said. The market share of RTP globally is about 4 percent with the RTP share of the U.S. market at roughly 15 percent; U.S. parks as a group have about 24 percent of the world market share, and International Association of Science Parks (IASP) member parks have about 72 percent of global share. "So we are not the center of the universe."

The park's university connections are "rich and robust," and each of the three university partners is involved in governance, leadership, and helping set strategy. The park employs some 40,000 full-time workers in 24.5 million square feet of developed space. The economic impact is $2.8 billion in capital investment and $2.7 billion in annual payroll. When the park began operation, about 11 percent of the employment in the region was in new-line or high-technology industries, and today this proportion exceeds 50 percent.

---

[17]Founded in 1959, Research Triangle Park is the nation's second oldest research park. Stanford Research Park—which was founded in 1951 as Stanford Industrial Park in Palo Alto, California—lays claim to be the world's first technology-focused office park.

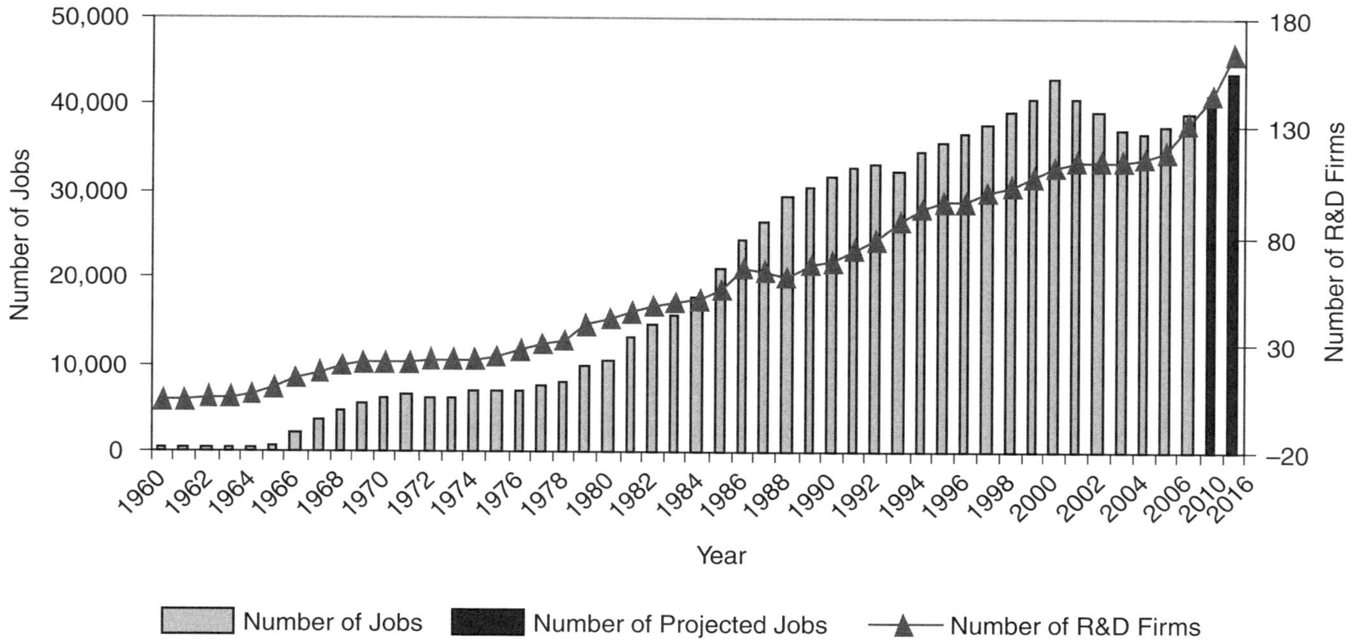

FIGURE 4 RTP's growth trajectory.

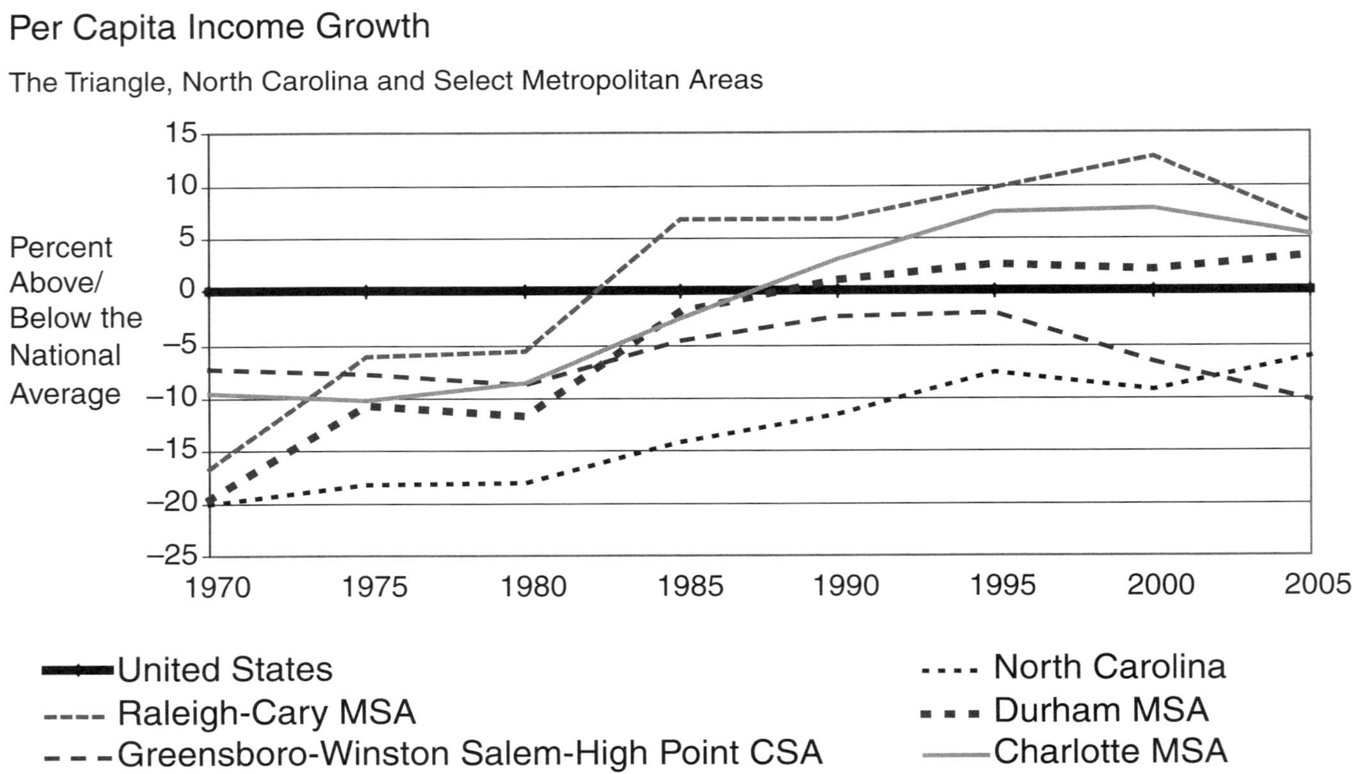

FIGURE 5 Long-term impact.
SOURCE: U.S. Bureau of Economic Analysis.

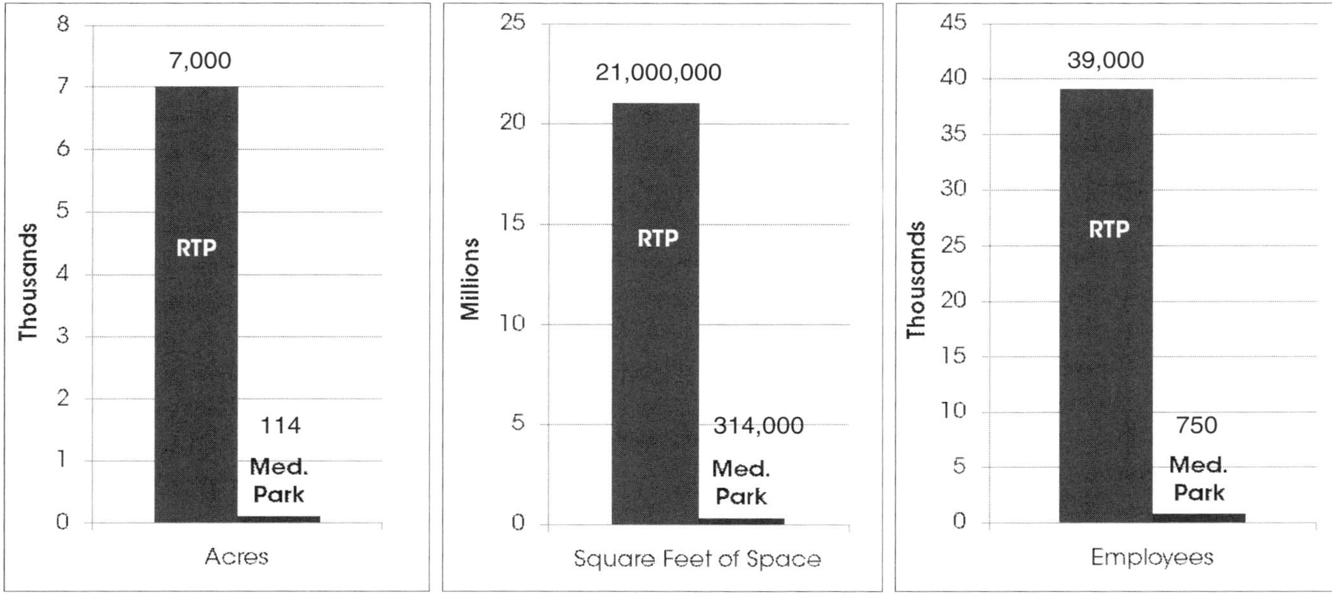

FIGURE 6 Size matters: RTP vs. other parks.
SOURCE: *Characteristics and Trends in North American Research Parks: 21st Century Directions*, Prepared by Battelle, October 2007.
NOTE: The "Typical North American Research Park" is defined here as the median research park.

A Diverse Mix of Industries

Mr. Weddle said that a strength of the park is its diverse industry mix: 29 percent of tenants specialize in life science, 21 percent in information technology, 13 percent in materials science and engineering, 15 percent in business and professional services, and 11 percent in scientific associations, foundations, and institutes. It was not always so: the early park was dominated by a few large companies. Between 1997 and 2007, the number of companies increased three-fold, with the number employing fewer than 250 employees rising from 53 to 150. Very large firms, employing more than 1,000, stayed relatively constant, with "good growth" in the 500-1,000-employee firms, "still our bread and butter." He said he is pleased with the movement into smaller companies.

One impact of the park has been the creation of some 1,500 start-ups by RTP companies and the three research universities since 1970. There are also four active incubators: Park Research Center, owned by the park; First Flight

Distribution of RTP Companies
(by number of employees)

Number of Employees	1997	2007
> 10,000	1	1
5,000 - 10,000	1	1
1,000 - 5,000	5	7
500 - 1,000	1	5
250 - 500	9	8
< 250	53	150
Number of Companies	70	172

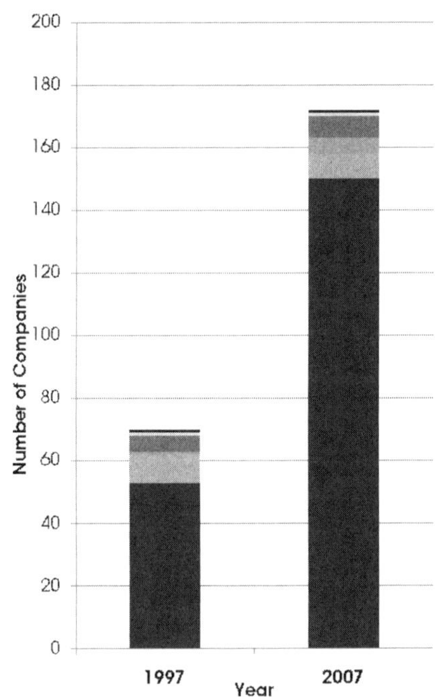

FIGURE 7 Diversity matters.

Venture Center, a successful nonprofit; Becton Dickenson BioVenture Center; and Alexandria Innovation Center.

Another impact of the park is "induced development." In analyzing all projects within a four-mile radius, the park found that in addition to 24.5 million square feet of built space, it has induced another 13 million square feet of built space. In addition, a great deal of residential development was transforming the landscape.

Expected Challenges

He listed several challenges expected to come from local and global changes:

- Changing nature of science: Parks must be ready for new trends.
- Shifts in corporate and government R&D: For example, Singapore's strategy includes large investments of public R&D money.
- Changing nature of work and the workforce: Of 10,000 employees at IBM, 40 percent of workers tele-work and do not come to the main campus in RTP every day. GlaxoSmithKline, with 6,000 employees, is "reconcentrating" its workers to spur innovation. "How can you be virtual and at the same time create face-to-face opportunities for synergies?"
- Sustainability issues: Energy, climate change, the green movement.
- Evolution of science parks as places to both work and live.
- Endless need for infrastructure investments: "When you're successful, traffic problems are your reward."

Mr. Weddle concluded by pointing out how important parks have become around the world. He said that while every park will have its own local challenges and mission, speakers at the symposium from diverse backgrounds had demonstrated the common need for parks in every country and the value of sharing best practices with others.

PURDUE RESEARCH PARK

Victor Lechtenberg
Purdue University

Purdue was established in 1869 as a land-grant university in Indiana.[18] Today

[18]The nation's land-grant colleges were authorized under the Morrill Act, signed into law in 1862 by President Lincoln, "to teach such branches of learning as are related to agriculture and the mechanic arts...." With very few exceptions, almost all of the 106 land-grant colleges are public, and include

it has about 39,000 students and 2,100 faculty. Dr. Lechtenberg noted that the Purdue University System—which includes its regional campuses statewide—is one of the largest public universities in the United States, with a reputation as a top engineering school.

Purdue University has a three-part mission of "learning, discovery, and engagement." It also has an economic development goal—to "advance Indiana's economic prosperity and quality of life." The Purdue Research Park supports this goal in several ways, through two separate legal entities: The Purdue Research Foundation and the Discovery Park.

Purdue Research Foundation: Parent of the Park System

The first, the Purdue Research Foundation, initiated a decade ago is situated on 725 acres near the main campus in West Lafayette, Indiana. The park missions are to train and empower students and faculty to develop and commercialize technologies needed to help the Indiana economy. Park facilities include use of the Internet, specialized laboratories, professional conference facilities and management, and access to the technical equipment of Purdue University.

The features of the Purdue Research Park include:

- The largest technology incubation program in the United States (259,000 square feet).
- 147 companies.
- 57 incubator businesses.
- Nearly 100 high-tech firms and entities.
- 2,800 employees.
- An average wage of $58,406 PTC.
- $121 million of invested venture capital.
- 52 separate facilities.
- 1.5 million+ square feet under roof.

The Purdue Research Park is the state's first Certified Technology Park.[19]

31 tribal colleges added in 1994. Some, including Purdue University, have developed far beyond their original mandate to include full arts and sciences curricula.

[19]The goal of Indiana's 2002 technology-park program is to boost high-tech economic development within specific areas identified by local development organizations. The program allows for the increased tax revenues generated by park tenants—including property, sales, and income taxes—to be reinvested into the park. The money can be used for lot improvements, facility operation and maintenance, payment on bonds, and other promotional activities. The program also offers grants of up to $500,000 to help get park development rolling. A host of requirements must be met before the state will declare an area a certified technology park. There must be significant support promised by a university and a commitment to the commercialization of products. A business incubator must be part of the plan, and local officials must line up at least one commitment by a high-tech company to

There are now 18 such certified parks distributed all over the state, each of which must have a university partner; Purdue is the university partner for 10 or 11 certified parks. Of these, half a dozen are doing quite well, said Dr. Lechtenberg; others are showing potential. The state did not put much money into the expansion but created a financing authority based on sales tax and income tax.

Discovery Park: Integrated Centers

Started in 2002, Discovery Park is a network of integrated research centers that are dedicated to large-scale, interdisciplinary research. Discovery Park has 11 *core centers* that are expected to have long or even indefinite life spans and focus on large research opportunities in biosciences, nanotechnology, advanced manufacturing, energy, oncology, and healthcare engineering. Discovery Park also has *project-based centers*. These are expected to be interdisciplinary in nature, sponsored by funders and affiliated with a core center. Their research is often based on emerging funding opportunities, including those in homeland security, climate change, visualization, and analytics transportation. New initiatives are being discussed, including cancer care engineering, cytomics, air emission monitoring, coal technology, and hydraulics.

Since its founding in 2002, funding for Discovery Park has grown to more than $50 million last year, and now represents a significant portion of the university's research portfolio. The Lilly Endowment has been instrumental in providing initial momentum.

A major objective of Discovery Park is to seed and nurture start-up companies. It has facilitated the start of 24 firms so far, as well as six start-ups initiated by students. This part of the program is called Interns for Indiana, another Lilly Endowment-funded project.

Indiana's Technical Assistance Program

Dr. Lechtenberg closed by describing the Technical Assistance Program (TAP), which achieved a statewide reach in 2007. During that year 150 TAP faculty, students, and staff worked in 413 companies, hospitals, and other institutions. The program develops consulting arrangements between universities and companies. Most of these have been in manufacturing fields until a few years ago, when a healthcare technical assistance program was added; last year a pharmacy TAP was begun. By mid-July 2008 the TAP will have offices at 12 locations around the state. The economic impact of this program on a small state has been significant, said Dr. Lechtenberg, with a cash value of over $80 million in increased sales, sales retained, cost savings, and capital investment. More than

---

operate within the park. See Steve Kaelble, "Good Neighbors: Indiana's Certified-Technology-Park Program," *Indiana Business Magazine,* July 1, 2004.

2,700 people have gained advanced manufacturing certificates in the past year, and more received workforce training, often in very advanced skills.

DISCUSSION

In response to a question, Dr. Lechtenberg said that Indiana's Technical Assistance Program is funded by a state appropriation that has not been increased since the mid-1980s. Program growth has come mostly in fees for services, contracts, and training programs.

A questioner asked about park size, and Mr. Weddle said that the park measured about two by eight miles. "This makes for difficult collaboration from company to company, and between our owners and the tenants association, both operationally and around communities of interest." He said that there is no one way to organize employees or incentives to improve interaction, but they are trying many approaches. The most popular park program has proven to be the softball league, with shortage of fields and a waiting list to enter a team. They also have other programs to encourage cross-company collaboration, including monthly mixers for the tech workers that are catered by restaurants from around the region. "We once hired someone to do salsa dancing lessons. Have you ever seen 300 people with pocket protectors doing salsa dancing?" Groups also formed online, including a wiki group, and task forces were set up to share best practices. "We're just beginning to dialogue with other parks," he said. "We're all collaborating, working hard to find ways to reduce the sensitivity around basic competition, and learning to work together."

On the topic of collaboration among institutions, he said that it is a challenging issue. Mr. Weddle of RTP said that the three universities "worked better with each other than most institutions work within themselves." He said that this collaboration was embedded early, when the state would fund facilities—such as the first supercomputer in the region—only if they were shared. When the biotech center was established in 1984, a single center had to be shared by the universities. "It is an institutionalized level of collaboration," he said. "The beneficiaries of our foundation are the three universities. We have a mechanism whereby we use our proceeds to fund demonstration projects that all agree on."

Dr. Lechtenberg of Purdue noted the challenge of assembling a critical mass of scientists and engineers on the smaller campuses. "What will be the emphases in smaller communities?" he asked. "The challenge is not in finding capital, but in finding the entrepreneurial talent locally."

The Evaluation Challenge and Policy Synergies

Moderator:
James Turner
U.S. House Science Committee

Mr. Turner began by acknowledging Dr. Mary Good as a mentor when they both worked for Signal Companies, before it became Allied Signal and was planning a relationship with Oak Ridge National Laboratory. Since then, he said, he has spent almost 30 years studying how federal legislation could promote innovation. He had worked on the Bayh-Dole acts, the Technology Transfer Act, various antitrust laws, and each reauthorization of the Small Business Innovation Research program, and other issues that set the framework for topics being discussed at the symposium.

With this background, he said, he had been thinking about how the world of start-ups and the kind of people who go into science and technology parks had changed. The federal government, he said, had paid little attention to the challenge of designing policies to encourage, plan, or evaluate research parks. The symposium has illustrated the energy with which other governments have made this issue a national priority, and he urged more attention domestically on how best to stimulate innovation at a time when the economy is a central issue.

THE NEED TO UNDERSTAND AND NURTURE INNOVATION

Mr. Turner said that he saw two urgent reasons to pay more attention to innovation. First, the United States must understand what is happening at the cutting edge of science, technology, and innovation around the world if it is to revive its economy and maintain a standard of living similar to what we have today. He noted that the economic stimulus package then being designed in the Congress would be "neither the first nor the last," but it was not addressing the fundamental framework in which innovation can flourish. "When a new [administration] comes in," he said, "it will be obvious that this issue is an important one. Knowing what's happening at the cutting edge, and how we can magnify that by an order of magnitude over a period of time, will tell us whether we are going to turn this economy around and maintain a standard of living similar to what we have now. This is an incredibly important exercise we're going through."

The second reason to pay more attention to innovation, he said, is that the world has changed dramatically in the past 10 years. The Internet has allowed business, education, and government to function in a more distributed way, and has enabled unprecedented collaborations between institutions and sectors. This country, he said, is in a position to nurture the companies that are on the cutting edge of technological practice, and this can influence how well the U.S. economy can compete in the future. But little attention has been paid on how best to nurture companies, or how to evaluate the results of those policies.

He recalled the roots of his experience in this arena during the Carter administration when he took part in a policy review of innovation. That was the stimulus for the Bayh-Dole acts, he said, and many subsequent policies of technology transfer and antitrust policies—"until the Internet came along. What we have to do now is figure out how the world has changed, how to communicate better in new ways, how to mine the new data we have, and how to share it using a common language and standards. What will that mean for technology parks, universities, government? If we could have everyone on that same wavelength we would have a tremendously different way of doing business."

THE ROLE OF SBIR AND STATE AWARDS

Robert McMahan
State of North Carolina

Dr. McMahan said that he works as senior science advisor to the governor of North Carolina and executive director of the Board of Science and Technology, which he called "the nation's first state-level office of science and technology policy." He said that while he has no direct role at the Research Triangle Park, he lives adjacent to it and "represents a constituency for the research park—those who are responsible for its funding and its promotion at the state level."

He said he would "take a somewhat contrarian view" in the discussion. While most of the participants have focused on inputs to science and technology parks, he said that he would focus also on outcomes and outcome measurements. He said he would use the state of North Carolina to illustrate the complex and sometimes counter-intuitive relationship of outcomes to inputs, particularly in the case of university research parks.

Transforming the Region

North Carolina, he said, has been recognized as successful in founding a research park that influenced the economic transformation of its region. Over 50 years ago, a small group of planners began to create structures to leverage innovation resources within the state. At the time, in the 1950s, North Carolina was "desperately poor," ranking 49th in the country in per capita income. Most of its population was involved in a few low-wage industries: tobacco, furniture, and textiles.

A few strong and committed leaders from the universities and the private sector came together to take action. The key question they asked, he said, is the same one people ask today: How do we leverage local innovation strengths to diversify and strengthen our economy? The idea of a science and technology park was then a novel one, but it made sense to these leaders. Planning quickly began through informal discussions in living rooms and offices. The original fundraising was done by a single individual named Archie Davis, who worked at Wachovia Bank. In 1958, Davis raised $2 million from private donations from across the state in 60 days—today's equivalent of $15 million—which made it possible to create a large and diverse innovation center for North Carolina.

The Importance of "Patient Structures"

What is distinctive about North Carolina's approach is that it recognized the importance of "patient structures," said Dr. McMahan. For the first 10 years, the park made little progress. But bridging institutions designed to institutionalize change, such as the Board of Science and Technology, provided continuing support. That Board also created innovations that have endured and served as models, especially the clusters of research, development, and innovation activities that "contribute enormous economic impact." The Board, with a long-term mandate, has been able to look ahead far enough to new fields gaining momentum, and urged the state to invest in biotechnology when biotechnology was not yet a familiar term. The biotech cluster is now the third largest such cluster in the nation. RTP, he said, used a planned approach that "differs from other parks that have grown in more organic fashion," he said, "and underscores the point that states are emerging as key drivers of collaborative R&D."

When looking at the inputs and outputs of the park, however, he reported one significant discrepancy. The state's investments have been strongly effective in supporting basic research and development and in the creation of innovation capacity. However, he said, it turned out that this high level of academic research support and performance is not matched by the strength of industrial R&D. So in terms of broad measures of commercialization and the total number of R&D dollars being spent, the state is somewhere around the U.S. average, or even in the "second tier of states."

The Challenge of Translating Research into "Gazelle" Firms

Dr. McMahan said that a closer study of this imbalance is even more revealing. In the United States as a whole, industry dominates R&D spending; for every $15 spent on industrial research and development, about $5 are spent on university R&D. The industrial spending is essential, he said, because it is one of the most reliable metrics for predicting the dynamism of the technology economy. While academic R&D creates innovation capacity, it does not by itself raise economic output or create the fast-growing "gazelle" firms that spur new growth. This has held true for RTP. While RTP ranks among the top parks in the country in innovation capacity, it ranks 48th in its creation of "gazelle firms"—small businesses with year-over-year growth above 20 percent for four consecutive years. "We see this same trend in emerging domains like nanotechnology," he said. "We are in the top ten in terms of nanotechnology research, but not in the number of companies being created."

It has been generally assumed, he said, that simply increasing levels of R&D proximate to universities raises the level of economic activity. "But often when you delve into the numbers this is not so. We need to understand structurally what the economic outcome of R&D is in terms of the major drivers of job growth and wealth creation in our economy." In the United States, a majority of jobs are created by small businesses; of all companies, he said, about 15 percent are responsible for the bulk of job creation. "So the importance of targeting economic growth in that segment is clear." RTP is now studying how better to convert its innovation capacity into jobs and other economic outcomes. One such effort is called UNC Tomorrow, a comprehensive plan to engage all 16 campuses of the state university system more closely around economic engagement.

A Commitment to SBIR

Dr. McMahan then came to the topic of SBIR, the Small Business Innovation Research program, which targets specifically the high-growth, high-potential companies mostly likely to achieve commercialization.[20] Statewide SBIR activity, he said, is an excellent proxy for the dynamism of the "gazelle" population.

For this reason, three years ago North Carolina instituted one of the country's most substantial SBIR support programs as part of its support of entrepreneurship.

[20]The National Academies has recently concluded the first comprehensive assessment of the program since its founding in 1982. See National Research Council, *An Assessment of the SBIR Program,* op. cit.

The state decided to award up to $100,000 in matching funds to each SBIR company. Already this approach has succeeded in targeting high-potential small firms, he said, "and should be viewed as a powerful complement of research parks and other structures."

He summarized by reiterating that the simple creation of parks and other organizations is not enough to sustain economic growth and create high-paying jobs. "We have to place as much weight on the environment in which they are created and the policy framework in which they exist. Otherwise, university-based regional economic activity is insufficient." This conclusion, he said, "drives us to a policy view of the university-based park as part of a continuum of mixed infrastructure and other measures. We have to ask ourselves constantly, how do we enhance their productivity and success? Entrepreneurship becomes critical to the success of the parks and the larger economy."

THE EVALUATION CHALLENGE

Albert N. Link
University of North Carolina at Greensboro

Dr. Link, a professor of economics who has long studied research parks in the context of economic growth, said he would offer his perspective on the challenge of evaluating park performance. He said that in 1985 he was asked to begin writing the history of Research Triangle Park, which has taken two volumes to compete. The first volume assembles an oral history, and the second is an account of the park's history and development.[21]

The Need for Accountability

Dr. Link noted two aspects to the challenge of evaluating a park such as RTP. The first is to identify the "environments" affected by park activities. The second is to identify evaluation metrics that are appropriate to each activity. Once the impacts have been identified, the review must design ways to quantify them and their costs.

He began with the question of why parks should be evaluated. His response was again twofold:

[21]A. N. Link, *A Generosity of Spirit: The Early History of the Research Triangle Park*, Research Triangle Park, NC: University of North Carolina Press for the Research Triangle Park Foundation, 1995; and A. N. Link, *From Seed to Harvest: The Growth of the Research Triangle Park*, Research Triangle Park, NC: University of North Carolina Press for the Research Triangle Park Foundation, 2002.

- Funds are needed for the development of park land and infrastructure, and more than 80 percent of parks rely on government or university funds or both.
- It is reasonable to expect that both "investors" and the local community will want accurate evaluations of parks.

This strong need for accountability, he said, is reasonable, because park directors are stewards of public money, and the public sector has a long history of being accountable for its use of public resources. This history dates back to the Budget and Accounting Act of 1921, which was followed by fundamental budget reforms carried out under President Wilson. More recently, Congress passed the Government Performance and Results Act of 1993 to update the terms of accountability for federal agencies.

Rationale for Evaluating Parks

Accountability is an especially important issue for universities, which, he said "are not known to be good managers." In this context, he suggested that exercises in evaluation actually help universities become better managers by helping them identify all the agents of the university that are affected by the park. In addition, he said, the budget constraints of the current period imply that the legislative initiatives that have supported the creation of parks will have to be justified, which will require at a minimum some in-depth case evaluations of selected parks around the country. "Many individuals in the administration today," he said, "who are affected by the decline in science budgets are suggesting that this is going to set science back quite a few years." One reason for this, he said, is that the need for the National Institutes of Health and the National Science Foundation is more apparent than the need of some other science-related activities, and these more visible and popular programs are more likely to grow. Another rationale for park evaluation, however, is that such an exercise often reveals ways to increase the efficiency of the science that takes place, both at the university and at independent companies.

The Government Performance Results Act (GPRA), passed in 1993 by Congress, has had a demonstrable trickle-down effect, he said.[22] States began to examine their own procedures and look for ways to quantify accountability. The universities, he said, also paid closer attention to what faculty do with their time, how much contact they really have with their students, and other activities. The trend toward evaluation is also going to trickle down to the science parks, he said,

[22]The Government Performance and Results Act of 1993 required federal agencies to develop long-term Strategic Plans defining general goals and objectives for their programs, to develop Annual Performance Plans specifying measurable performance goals for all of the program activities in their budgets, and to publish an Annual Performance Report showing actual results compared to each annual performance goal.

because a good deal of public money goes into creating and maintaining them. He pointed out that many European countries evaluate their parks even before they are funded, as well as afterward.

Many Kinds of Outcomes

He suggested many kinds of outcomes that can justify investments in parks. Among the more apparent are spillovers to the global economy that "usually take the form of the creation of codified knowledge." Spillovers, he said, can also be examined in other ways. Tenants often form research joint ventures with other firms in the park. Also, tenant companies may throw off benefits to the host university by sponsoring laboratories and professorships, hiring students, or associating themselves with co-patenting activity. (He noted a shortcoming in co-patenting activities, in that it generally results in the patent going "out the back door from the university directly into the company.") Finally, the local community may induce in-migration of other companies into the area, some of which may locate in the park.

Another kind of outcome is the value of the park to tenant companies, which benefit from the richness of the flow of knowledge between them and universities. "The greatest challenge is to keep them in the park for extended periods," he said. Although it is not common for one park to lure tenants away from another, companies may cut back on their level of R&D, or consolidate their research organizations at a central laboratory. The best way to keep tenants is to ensure that the flow of knowledge continues at a high rate. "This," he said, "is a challenge for the university and the reward structure of the university."

How Universities Benefit from Parks

While parks gain knowledge and prestige from their university partners, the universities gain as well. Association with the park brings the university higher publication rates, more successful patenting activities, greater ability to hire eminent scientists, and larger extramural grants. Interviews with provosts of universities associated with parks say that one result is to shift curricula toward more applied research and to improve the ability to hire graduate students in science and engineering. The university as a whole becomes more interested in innovation. Faculty who mentor students are more attuned to the needs of industry. The change in the structure of the curriculum is reflected by improved student placements.

The "Cachet" of a Park

Another outcome that may be less obvious, he said, is the cachet of working in a successful park, which can benefit the host university, tenant firms, and the local community. He recalled the example of the Semiconductor Research Corporation (SRC), a consortium of U.S. and now international semiconductor firms formed primarily to work out intellectual property agreements with many universities. It was initially located in RTP, but as it began operation, some members were more satisfied than others. A few companies talked about leaving, saying that the park was too expensive. In response, the SRC cut costs by moving out of the park. At the same time, it wanted to maintain the "halo effect" of a park address, so it kept its mailbox at RTP.

There is a need, he said, for more studies that compare the performances of firms located in parks with those of comparable firms not in parks. He said that the best study, done in the UK, found that the performance of firms is greatly enhanced by location in a park and juxtaposition to a university.

Rewarding Those Who Create Knowledge

Dr. Link closed by offering several reflections on park evaluations. For one thing, he said, identifying the stakeholders and the many linkages between them is "easier said than done." University administrators have a tendency to recognize their own "outward" influence on job creation and other impacts on the local community. But they are less likely to ensure that faculty who create this flow of knowledge between university and firm are rewarded and appreciated.

He concluded by predicting that more parks are coming, but that they will need better ways to demonstrate their value. "There are about 100-plus parks that have a formal relationship with a university," he concluded, "and about another 25 to 30 in planning stage. I think these numbers will grow; not automatically—not 'if we build it they will come.' That model applied in the early 1990s but not now. These new parks are going to depend on placement near specific universities and strength in specific technologies."

DISCUSSANT

William Kittredge
U.S. Department of Commerce

Dr. Kittredge, who heads the Economic Development Administration's Office of National Programs and Performance Evaluation (NPPE) within the Department of Commerce, said that his office performs several functions relevant to the evaluation of S&T parks. NPPE reviews all proposed investments—such as the fiber-optic system recently installed at Sandia—as a pre-audit function and, in doing so, seeks to advance the science of measurement and outcome prediction. NPPE also performs evaluations retrospectively to assess the outcomes of federal investments. Dr. Kittredge serves in a similar capacity as U.S. representative to the Organisation for Economic Cooperation and Development (OECD).

He said that developing metrics to measure the performance of S&T parks and of economic development in general is a "work in progress." What we do know, he said, is that people want to evaluate outcomes. They want to know "whether they, their children, and their community are going to prosper and be competitive in the global economy." But no measurement system will be able to determine policy in this area, because so much of the decision making is "socially mediated." What is successful for the French, he said, may not work for the United States. He also stated his opinion that while outcome measurement informs public policy decisions, it is not dispositive.

The Limits of Quantitative Measures

With respect to the yearning for quantitative measurements, he quoted a former professor who said, "If you can't count it, it doesn't count." With due respect to the professor, he said, it is apparent that qualitative measures also have an important role in park evaluation because qualitative measures add dimensionality and nuance to the more generalized quantitative measurements. The result is a more complete understanding of the characteristics of successful parks enhancing our ability to make fruitful science and technology park investments.

Measurement limitations are also imposed by definitional limitations, he said. More accurate evaluation awaits broader agreement on a definition of innovation and other key terms.

"How do we decide what to accept as metrics?" he asked. For business incubators, it seems natural to measure the private investment and job creation. However, he gave one example of how unreliable this can be. He was asked to evaluate business incubators the department had supported. He found that one particular incubator, six years after it began operation, had listed no jobs and no private investment. What had happened, he said, was that by the time they did their assessment, both the companies had already gone public, so "nobody was there. That was actually a good thing. We need to be careful about what we measure, how we measure, and when we measure" to effect accurate evaluation of science and technology park investments. "Multi-dimensional evaluation that includes private investment, graduation rate, and other key variables is the key to meaningful evaluation and credible communication of our results," he said.

He added that one has to be careful about expectations. In a research park, one might expert to measure success in terms of either serial entrepreneurship or a long-term park presence, whereas a firm in an incubator might be more interested in "graduating" from the park. "Whatever we decide," he said, "we need a common language for the business community, investors, university, researchers, engineers, and other stakeholders."

Some Qualities of Good Metrics

Quantitative measures are important, he said, because "I want something I can see." But they are only one kind of tool. He said that it would be impossible to predict the success of an innovative firm such as Federal Express by using traditional quantitative measures. "Measures need to have certain characteristics," he said. "They have to be credible and make claims that can be sustained. They have to be transparent, material, contextually important. We have to be able look at our failures, assess them, and accept that it was experimentation that didn't work. I personally object to the 'but for' justification: If we hadn't done this, this wouldn't be here. Well, you don't know that for sure and making the claim undermines our credibility."

"Good" metrics evolve as our expectations and aspirations grow and change. "I understand that some people in the Research Triangle are now focusing on value-added activities that grow from the world-class area we all recognize as a model of success." The new metrics under consideration seek to retain more value-added activity in the Research Triangle area rather than exporting innovations to be implemented elsewhere.

In summary, he suggested that a good outcome measure is one that demonstrates the value proposition for stakeholders, and that we are far from having such a measure. "We owe the stakeholders a credible review of what their investment is doing. But I don't think we're going to find a single metric for that. Our goal is to increase our knowledge as we go along, knowing that it's difficult to get cause and effect together. We can get improvements, so we can become more credible; we can understand what good investments are, make more of them, move forward, and expand them. Forums like this are important steps."

Closing Remarks

Mary Good
University of Arkansas at Little Rock;
and Board on Science, Technology, and Economic Policy

Dr. Good began by recalling how different this conversation might have been 10 years ago, when S&T parks were focusing more on domestic competition. "Today," she said, "we have heard from a wide variety of places and circumstances. We now understand that what is going on outside our borders is also important and that we need to understand it. We are making progress. It's slow, but little by little we are learning."

COMMON THEMES: ATTRACT AND TRAIN HIGH-QUALITY PEOPLE

She said that much of what was said during the day is particular to individual parks, but that she heard at least two common themes. One is that successful parks must begin with sustained leadership backed by high-quality personnel. The second is that successful parks create new high-quality jobs for the future. "Whether we want to look at the goals of parks in those terms," she said, "that's what the people here today were talking about. If you have not recruited the right people, and increased their number for the future, the research park is likely to be a failure, no matter where the parks are or what the external issues are."

AN UNDERCURRENT OF URGENCY

She said she also detected a change of pace in the planning and development of parks around the world. Time and patience were important, she agreed, "but we have heard an undercurrent of urgency from essentially all of them. Sovereign states have decided for the most part to front-load economic development and to lean on their state-supported universities, even though state support is not what it used to be. Research parks are clearly a part of economic development today, and if we can look at what we've learned here today and disseminate it quickly, we can do some good."

The discussion on assessment is particularly timely and necessary, she said, because "all of you associated with research parks are going to have to figure out what to do with respect to assessment. The materials from this meeting can certainly be a part of that."

PARKS AND ECONOMIC DEVELOPMENT

Dr. Good closed by thanking all participants, and especially the speakers, for being "candid, thoughtful, and insightful." She praised the content of the meeting for its high quality, and said that "with any luck it will be applicable to all of us and beyond as we grapple with the major issues of economic development, for that is what parks are surely about."

She said that parks can and must flourish beyond the few dominant centers of science and technology expertise. "This is a message you can take home to your governors and state legislatures," she said. "These issues need to be getting a good hearing in all the states these days, and they're not. If we don't succeed in supporting innovation around more of our universities throughout the country, the United States could easily end up with lots of activity on both coasts and a lot of flyover country in between. If that happens, we as a nation are not going to be as visible or viable in the next 50 years as we have been in the past."

III

RESEARCH PAPER

Research, Science, and Technology Parks: An Overview of the Academic Literature

Albert N. Link
Department of Economics
University of North Carolina at Greensboro

I. INTRODUCTION

Research, science, and technology parks are:

> … seen increasingly around the world as a means to create dynamic clusters that accelerate economic growth and international competitiveness (Introduction to the report, p. 2)

As such, it is important to understand the academic literature related to research, science, and technology parks (hereafter R-S-T parks, or simply parks) because that literature, albeit embryonic, has had and will continue to frame public policies related to park formations and growth. The purpose of this background paper is thus to overview the extant academic literature and to anticipate public policy issues that have not yet been debated.[1]

The remainder of this paper is outlined as follows. In Section II, dimensions of a definition of a park are set forth. In Section III, alternative theories on R-S-T park formations are overviewed. Section IV summarizes the extant empirical literature and places it in the context of a model of innovation. The paper concludes with policy considerations in Section V.

II. DEFINITIONS

The term research park is more prevalent in the United States, the term science park is more prevalent in Europe, and the term technology park is more

[1]This paper draws on Link and Scott (2007).

prevalent in Asia. Many definitions of a park have been proffered, mostly by professional organizations (e.g., AURP 1998, IASP 2000, UKSPA 2003, and UNESCO 2004) and by parks themselves as a way to define their activities. Common among these definitions is that a park is a type of public-private partnership that fosters knowledge flows—often between park firms and universities and among park firms—and contributes to regional economic growth and development.[2]

Link and Scott (2006), based on an overview of alternative definitions of university research parks, and most parks in the United States are affiliated with a university, propose the following definition:

> A university research park is a cluster of technology-based organizations that locate on or near a university campus in order to benefit from the university's knowledge base and ongoing research. The university not only transfers knowledge but expects to develop knowledge more effectively given the association with the tenants in the research park.

A public-private partnership, with reference to a park, is an infrastructure that leverages, formally or informally, the efficiency of innovation that takes place within park firms and within universities, when present. "Public" refers to any aspect of the innovation process that involves the use of governmental resources, be they federal or national, state, or local in origin. "Private" refers to any aspect of the innovation process that involves the use of private-sector resources, mostly firm-specific resources. And, resources are broadly defined to include all resources—financial resources, infrastructural resources, research resources, and the like—that affect the general environments in which innovation occurs. Finally, the term "partnership" refers to any and all innovation-related relationships, including but not limited to formal and informal collaborations in R&D.

In the case of parks in the United States, government involvement tends to be indirect with economic objectives of leveraging public-sector R&D (including university R&D) and private-sector R&D. In many Asian countries, for example, government involvement is direct rather than indirect.[3]

III. THEORIES ON R-S-T PARK FORMATIONS

Surprisingly, the extant literature in economics, geography, management, and public policy does not offer a fully developed theory about the formation of parks. Case studies have documented the institutional history of a number of research parks, university affiliated or not. Castells and Hall (1994) describe

---

[2]These definitional characteristics are emphasized by President Mote of the University of Maryland and President Barker of Clemson University in this report.

[3]Direct government involvement in park activity is illustrated through the many summaries of activities in Asian parks in this report.

the Silicone Valley (California) and Route 128 (around Boston, Massachusetts) phenomenon; Luger and Goldstein (1991), Link (1995, 2002), and Link and Scott (2003a) detail the history of Research Triangle Park (North Carolina); Gibb (1985), Grayson (1993), Guy (1996a, 1996b), and Vedovello (1997) summarize aspects of the science park phenomenon in the United Kingdom; Gibb (1985) also chronicles the science/technology park phenomenon in Germany, Italy, Netherlands, and selected Asian countries; and Chordà (1996) reports on French science parks, Phillimore (1999) on Australian science parks, Bakouros et al. (2002) and Sofouli and Vonortas (2007) on the development of science parks in Greece, and Vaidyanathan (2008) on technology parks in India.

Scholars have not yet formally tied the emergence of parks to cluster theory, although cluster theory has been applied to the formation of biotechnology and other science-based agglomerations of firms near universities so the potential application is not unreasonable. Drawing on cluster theory—and location theory was, in part, a prequel to the popularization of cluster theory, as reviewed by Goldstein and Luger (1992) and Westhead and Batstone (1998)—one could argue that there are both demand and supply forces at work that result in the clustering of research firms near universities (Baptista, 1998).

On the demand side, there are sophisticated users of developed technologies within a park, and the search costs for such users are minimized by locating on a park. Of course, there are disadvantages associated with being in a park, mainly greater competition for the developed technologies. On the supply side, there is skilled and specialized labor available from the university or universities involved in the park in the form of graduate students and consulting faculty, although there is also more competition for that pool of human capital. Also, for a firm, location on a park, especially a university park, provides a greater opportunity for the acquisition of new knowledge—tacit knowledge in particular. As well, for the university, having juxtaposed firms provides a localized opportunity for licensing university-based innovations. The theory of agglomeration economics emphasizes knowledge spillovers and enhanced benefits and lowered costs caused by the presence of multiple organizations and the externalities they create (Swann, 1998). And, Audretsch (1998); Audretsch and Feldman (1996, 1999); Breschi and Lissoin (2001); Jaffe (1989); Jaffe, Trajtenberg, and Henderson (1993); and Rothaermel and Thursby (2005a, 2005b) provide empirical support for the agglomeration effect.

Henderson (1986) and Krugman (1991) emphasize conceptually as well as empirically the importance of location *per se* with regard to knowledge spillovers. Localization has an effect on resource prices. To the extent that new technology embodies new knowledge, geographic closeness implies lower new technology prices and thus presumably greater usage. Firms achieve economies of scale more easily with newer technologies. Arthur (1989) underscores the related importance of network externalities with regard to such scale economies. David (1985) also argues in general—and his argument could apply particularly well to university parks—that chance or historical events can lock a technology on a particular path of development. If that technology had a university origin, then creating such a park, from the university's perspective, and locating in the park, from a firm's perspective, gives positive feedback to continue the path dependency of the particular technology. The idea of path dependency, according to Arrow (2000), has its origins in the early writings of economists Veblen and Cournot, but it also can be traced to the Nelson and Winter (1982) concepts about evolutionary economics.

Relatedly, Leyden, Link, and Siegel (2008) outline a theoretical model, based on the theory of clubs, to describe the conditions under which a firm would be located in an existing university park. The authors conjecture that a university research park acts like a private organization, so that membership in the research park is the result of mutual agreement between the existing park tenants including the university, the club, and a potential new member firm. The decision to admit the new firm depends on the marginal effect of that firm on the wellbeing of the firms already in the park. For the representative in-park firm, the value of belonging to the park is the opportunity to engage in synergistic activities, which can be used to increase its profits in the output markets in which it participates, net of the direct cost (e.g., maintenance cost of being in the park and maintaining infrastructure) and indirect cost (e.g., congestion and competition for new knowledge) of being in the park.

IV. EMPIRICAL STUDIES OF R-S-T PARKS

The empirical literature related to parks is embryonic. Table 1 summarizes findings from the extant literature in four dimensions:

- Factors affecting firm decisions to locate on a park.
- Formation of university parks and university performance.
- Firm performance on a park.
- Parks and regional economic growth development.

It is clear from the literature review in Table 1 that R-S-T parks, especially university research parks, matter in several dimensions related to innovative activity and to economic growth and development.

To place the importance of these empirical findings in a broader perspective, consider the model of economic development in Figure 1. At the root of the model is the science base, referring to the accumulation of scientific and technological knowledge. The science base resides in the public domain. Investment in the science base comes through basic research, primarily funded by the government and primarily performed globally in universities and federal laboratories.

For an integrated technology-based manufacturing firm—and the model is similar for a service sector firm (Gallaher, Link, and Petrusa 2006) except that technology is generally purchased rather than induced through own R&D—technology development in the form of basic and applied research generally begins within its laboratory. There, R&D involves the application of scientific knowledge toward the proof of concept of a new technology. Such fundamental research, if successful, yields a prototype or generic technology. If the prototype technology has potential commercial value, follow-on applied research takes place followed by development. If successful, a proprietary technology will result. Basic research, applied research, and development occur within the firm as a result of its strategic planning and guide the firm's market-oriented entrepreneurial activities. Generally, strategic planning involves the formulation of road maps for developing new emerging technologies. A manufacturing firm targets discrete technology jumps, creating new technologies that make its competition obsolete; its strategic plans are long term and not closely linked to current competitive planning. Entrepreneurial activity then drives the firm toward the production of the new product or process. With entrepreneurial activity, a causal element in this model, the overall innovation process exhibits hysteresis because of the lagging impact of such enteperneurship. Thus, the relationship between investment in proprietary technology and market development might not be as predictable as the firm's strategic planners would like.

Infratechnologies (i.e., infrastructural technologies) emanate from the science base and from various technology infrastructures such as national laboratories. These technologies, such as test methods or measurement standards, reduce the market risk associated with the introduction of a new product or process. Once a new product has been designed and tested, technical risk could be low, but market risk could be significant until the product is accepted and adhibited and integrated into existing systems (e.g., in a service sector firm). The nonlinearity of this system is, literally, in the fact that there are multiple influences on both innovation and technology development, thus underscoring the existence of a need for broad-based and multitargeted public sector innovation and technology policy actions, not all of which are necessarily entrepreneurial.

The impact of R-S-T parks within this model is on the science base and on proprietary technology. Parks, university parks in particular, enrich the science base because they leverage the university's R&D. Parks also leverage firm's R&D, especially in comparison to off-park firms.

V. POLICY CONCLUSIONS

The elements of a national innovation system include competitive firms and a competitive environment, an effective educational system, strong university research, a legal system with property rights, and a capital market that includes venture capital (Nelson 1993, Cohen 2002). R-S-T parks have a unique place within a national innovation system.

Although the literature related to parks is still embryonic, the evidence

TABLE 1 Empirical Studies of Research, Science, and Technology Parks

Research Issue	Author(s)	Dimensions of Study	Findings
Factors affecting firm decisions to locate on a park	Westhead and Batstone (1998)	Comparison of UK on-park and off-park firms.	Location on a park is driven by the firm's desire to acquire research facilities and scientists at the university—all UK parks are located on or near a university.
	Goldstein and Luger (1992)	Comparison of university-based and nonuniversity-based parks in the United States.	Key criterion for location on a park is the linkage between the firm and the university (or, if generalizable to other countries, the higher education institution).
	Hansson, Husted, and Vestergaard (2005)	Case studies of UK and Denmark parks.	Firms locate in the park because of a need for social capital to facilitate entrepreneurial growth.
	Leyden, Link, and Siegel (2008)	U.S. public firms that have and do not have a research facility on a university park.	Parks invite firms to join a park based on their potential spillover benefits (i.e., knowledge spillover benefits) to existing park firms.
Formation of university parks and university performance	Link and Scott (2003b)	Growth of U.S. university parks over time.	Growth of park formations follows a Gompertz survival-time model; formal park-university relationships lead to increased university publication and patenting activity, greater extramural funding success, and enhanced ability to hire preeminent scholars.
Firm performance on a park	Westhead (1995), Westhead and Cowling (1995), Westhead and Storey (1994, 1997), and Westhead, Storey, and Cowling (1995)	Matched pair comparison of on-park and off-park UK firm performance.	Survival rate of on-park firms greater than of off-park firms.

	Siegel, Westhead, and Wright (2003)	Matched pair comparison of on-park and off-park UK firm performance.	Research productivity of on-park firms greater than of off-park firms.
	Lindelöf and Löfsten (2003, 2004)	Matched pair comparison of on-park and off-park Swedish firms.	On-park firms place greater emphasis on innovative ability, sales and employment growth, market orientation, and profitability than off-park firms.
	Ferguson and Olofsson (2004)	Matched pair comparison of on-park and off-park Swedish firms.	No performance differences between on-park and off-park firms.
	Fukugawa (2006)	Matched pair comparision of on-park and off-park Japanese firms.	Research linkages more likely formed with universities if on a park than off of a park.
Parks and regional economic growth and development	Goldstein and Luger (1992)	Descriptive analysis of U.S. park directors.	Parks leverage new business startups.
	Shearmur and Doloreux (2000)	Descriptive analysis of Canadian park directors.	Parks leverage new business startups and overall employment growth.

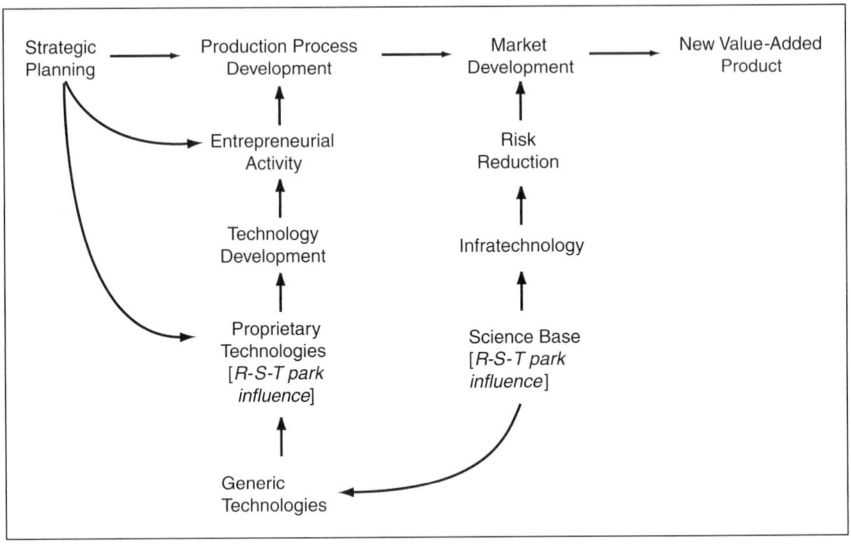

FIGURE 1 Model of innovation: A manufacturing firm.

suggests that parks enhance the two-way flow of knowledge between firms and between firms and universities, when a university is present. Thus, parks enhance innovation and subsequently competitiveness as was suggested in Figure 1.

Many nations' sectors have to varying degrees informally encouraged the formation of industry/university linkages. France's central government, like that of Japan, the Netherlands, and the United Kingdom, has actively fostered the creation of science parks (Westhead 1997, Hilpert and Ruffieux 1991, Goldstein and Luger 1990), and Germany has long promoted academic innovation centres to incubate and develop small- and medium-sized enterprises (Sternberg 1990).

In the United States, public investment at state universities is used to underwrite the formation and development of R-S-T parks. In 2004, through Senator Bingaman's introduction of S. 2737, "The Science Park Administration Act of 2004," and again in 2007 through Senator Pryor's introduction of S. 1373, "The Building a Stronger America Act," the U.S. Congress considered, but did not pass, a bill to provide grants and loans to states and local authorities for the development and construction of university parks. Implicit in these bills is the assumption that R-S-T parks are an important element in the U.S. national innovation system, and as such should be fostered because of both the knowledge-based and employment-based spillovers that will result.

This U.S. action may the most obvious example of public-sector support for R-S-T parks. Hand-in-hand with public-sector support is the need for public accountability, namely the development and implementation of evaluation methods and tools not only to support the assumption that R-S-T parks are in fact an important element of the national innovation system but also to quantify the net spillover benefits that result from public-sector support.[4]

The matched pairs studies discussed above in Table 1 are a preliminary form of evaluation. That is, it is useful to know that there is evidence that firms located on a park are more productive than firms not on a research park, ceteris paribus. However, when substantial public-sector resources are devoted to park formations, a more in-depth evaluation approach is warranted, namely the application of what Link and Scott (2001) call the spillover evaluation method. The spillover evaluation method applies to publicly funded, privately performed research projects, and research projects are defined in terms of the research activities that occur in the park rather than simply the construction of the park.[5]

There are important projects where economic performance can be improved with public R&D funding of privately performed research. Public R&D funding is needed when socially valuable projects (i.e., when the marginal benefits of the project are greater than the marginal costs) would not be undertaken without it.[6] If their expected rate of return from creating an R-S-T park environment falls short of their required rate, called the hurdle rate, then the university or local firms would not invest in the research park environment. Nonetheless, if the benefits of the research spill over to consumers and to firms other than the ones investing in the research, the social rate of return may exceed the appropriate hurdle rate, even though the private rate of return falls short of the private hurdle rate. It would then be socially valuable to have the investments made, but since the university or local firms will not make them without public support, the public sector should support the investments. By providing public funding, thereby reducing the investment needed from the university and local firms doing the research, the expected private rate of return can be increased above the hurdle rate. In this case, the public-sector's support may also suggest, or affirm, the possibility of a market for a successful project, thus reducing the investors' perceived risk as well as increasing the initial investment they are willing to make. Thus, because

---

[4]See presentation by Albert Link in this report.

[5]If one defined narrowly the output of the use of public-sector resources as the park itself, then, following Link and Scott (1998), the counterfactual evaluation method would be appropriate. When publicly funded, publicly performed research investments are evaluated, and the public is building the park, one should ask: What would the private sector have had to invest to achieve the benefits associated with the park in the absence of the public sector's investments? The answer to this question gives the benefits of the public's investments, namely, the costs avoided by the private sector.

[6]There is a rich history of economic and policy research to support this view. Notable are the arguments of Arrow (1962), Tassey (1997), Jaffe (1998), and Link and Scott (1998). Public funding of privately performed research is termed "public provision" while research conducted at wholly public facilities is "public production." Vincent Ostrom, Charles Tiebout, and Robert Warren distinguish between the production and provision of public goods. See V. Ostrom, C. M. Tiebout, R Warren, "The Organization of Government in Metropolitan Areas: A Theoretical Inqury," *American Political Science Review* 55, 1961.

of the public subsidy, the university, when present, and local firms are willing to perform the research that is socially desirable because much of its output spills over to other firms in the park and sectors in the local and national economy.

The question asked in the spillover evaluation method is one that facilitates an economic understanding of the potential returns to public-sector support for a portion of private-sector research, namely: What proportion of the total profit stream generated by the university's and local firms' research and innovation does the university and local firms expect to capture; and hence, what proportion is not appropriated but is instead captured by others that use knowledge generated by park research to produce competing products for the social good?[7]

As this overview shows, a case can be made that R-S-T parks are an element of a national innovation system, not necessarily a primary element but an important element none the less. Successful two-way knowledge flow between universities and industry is indeed a key ingredient for a national innovation system, and we do have evidence that R-S-T parks play a role in that knowledge flow. However, parks are not a *sine qua non* of the knowledge flow. Perhaps, consistent with the findings of the survey of university provosts reported in Link and Scott (2003b), R-S-T parks fall under the broader category of an effective educational system. However, R-S-T parks may in the future warrant a higher status, especially as technological life cycles continue to shorten and as basic research at universities (and to a growing extent at national laboratories [National Research Council, 1999, 2001]) and applied research/development in industry become more intertwined.

REFERENCES

Arrow, K. J. 1962. "Economic Welfare and the Allocation of Resources for Invention." In *The Rate and Direction of Inventive Activity*. Princeton, NJ: Princeton University Press.

Arrow K. J. 2000. "Increasing Returns: Historiographic Issues and Path Dependence." *European Journal of the History of Economic Thought* 7(2):171-180.

Arthur, W. 1989. "Competing Technologies, Increasing Returns, and Lock-in by Historical Small Events." *Economic Journal* 99(2):116-131.

Association of University Research Parks (AURP). 1998. "Worldwide Research & Science Park Directory 1998." New York: BPI Communications.

Audretsch, D. B. 1998. "Agglomeration and the Location of Innovative Activity." *Oxford Review of Economic Policy* 14(2):18-29.

[7]The part of the stream of expected profits captured by the innovator is its private return, while the entire stream is the lower bound on the social rate of return (because of the additional benefits of consumer surplus and assuming any cannibalization of existing surplus is relatively small). The spillovers evaluation weighs the private return (in practice—see Link and Scott [2001]—estimated through extensive interviews with the private-sector organizations receiving public support regarding their expectations of future patterns of events and future abilities to appropriate returns from R&D-based knowledge) against private investments. The social rate of return weighs the social returns against the social investments.

Audretsch, D. B., and M. P. Feldman. 1996. "R&D Spillovers and the Geography of Innovation and Production." *American Economic Review* 86(3):630-640.

Audretsch, D. B., and M. P. Feldman. 1999. "Innovation in Cities: Science-based Diversity, Specialization, and Localized Competition." *European Economic Review* 43(2):409-429.

Bakouros, Y. L., D. C. Mardas, and N. C. Varsakelis. 2002. "Science Parks, a High-Tech Fantasy? An Analysis of the Science Parks of Greece." *Technovation* 22(2):123-128.

Baptista, R. 1998. "Clusters, Innovation, and Growth: A Survey of the Literature." In G. M. P. Swann, M. Prevezer, and D. Stout, eds. *The Dynamics of Industrial Clustering*. Oxford, UK: Oxford University Press.

Breschi, S. and F. Lissoin. 2001. "Knowledge Spillovers and Local Innovation Systems: A Critical Survey." *Industrial and Corporate Change* 10(4):975-1005.

Castells, M., and P. Hall. 1994. *Technopoles of the World*. London, UK: Oxford University Press.

Chordà, I. M. 1996. "Towards the Maturity State: An Insight into the Performance of French Technopoles." *Technovation* 16(3):143-152.

Cohen, W. 2002. "Thoughts and Questions on Science Parks." Presented at the National Science Foundation Science Parks Indicators Workshop. University of North Carolina at Greensboro.

David, P. A. 1985. "Clio and the Economics of QWERTY." *American Economic Review* 75(2):332-337.

Ferguson, R. and C. Olofsson. 2004. "Science Parks and the Development of NTBFs: Location, Survival and Growth." *Journal of Technology Transfer* 29(1):5-17.

Fukugawa, N. 2006. "Science Parks in Japan and Their Value-Added Contributions to New Technology-based Firms." *International Journal of Industrial Organization* 24(2):381-400.

Gallaher, M. P., A. N. Link, and J. E. Petrusa. 2006. *Innovation in the U.S. Service Sector*. London, UK: Routledge.

Gibb, M. J. 1985. *Science Parks and Innovation Centres: Their Economic and Social Impact*. Amsterdam, The Netherlands: Elsevier.

Goldstein, H. A., and M. I. Luger. 1990. "Science/Technology Parks and Regional Development Theory." *Economic Development Quarterly* 4(1):64-78.

Goldstein, H. A., and M. I. Luger. 1992. "University-based Research Parks as a Rural Development Strategy." *Policy Studies Journal* 20(2):249-263.

Grayson, L. 1993. *Science Parks: An Experiment in High-Technology Transfer.* London, UK: The British Library Board.

Guy, I. 1996a. "A Look at Aston Science Park." *Technovation* 16(5):217-218.

Guy, I. 1996b. "New Ventures on an Ancient Campus." *Technovation* 16(6):269-270.

Hackett, S. M., and D. M. Dilts. 2004. "A Systematic Review of Business Incubation Research." *Journal of Technology Transfer* 29(1):55-82.

Hall, B. H., A. N. Link, and J. T. Scott. 2001. "Barriers Inhibiting Industry from Partnering with Universities: Evidence from the Advanced Technology Program." *Journal of Technology Transfer*. 26(1-2):87-98.

Hall, B. H., A. N. Link, and J. T. Scott. 2003. "Universities as Research Partners." *Review of Economics and Statistics* 85(2):485-491.

Hansson, F., K. Husted, and J. Vestergaard. 2005. "Second Generation Science Parks: From Structural Holes Jockeys to Social Capital Catalysts of the Knowledge Society." *Technovation* 25(9):1039-1049.

Henderson, J. V. 1986. "The Efficiency of Resource Usage and City Size." *Journal of Urban Economics* 19(1):47-70.

Hilpert, U., and B. Ruffieux. 1991. "Innovation, Politics and Regional Development: Technology Parks and Regional Participation in High-Technology in France and West Germany." In U. Hilpert, ed. *Regional Innovation and Decentralization: High-Technology Industry and Government* Policy. London, UK: Routledge.

International Association of Science Parks (IASP). 2000. <http://www.iaspworld.org/information/definitions.php>.

Jaffe, A. B. 1989. "Real Effects of Academic Research." *American Economic Review* 79(5): 957-970.

Jaffe, A. B. 1998. "The Importance of 'Spillovers' in the Policy Mission of the ATP." *Journal of Technology Transfer* 23(1):11-19.

Jaffe, A. B., M. Trajtenberg, and R. Henderson. 1993. "Geographic Localization of Knowledge Spillovers as Evidenced by Patent Citations." *Quarterly Journal of Economics* 108(3):577-598.

Krugman, P. 1991. *Geography and Trade*. Cambridge, MA: MIT Press.

Leyden, D. P., A. N. Link, and D. S. Siegel. 2008. "A Theoretical and Empirical Analysis of the Decision to Locate on a University Research Park." *IEEE Transactions on Engineering Management* 55(1):23-28.

Lindelöf, P., and H. Löfsten. 2003. "Science Park Location and New Technology-Based Firms in Sweden: Implications for Strategy and Performance." *Small Business Economics* 20(3):245-258.

Lindelöf, P. and H. Löfsten. 2004. "Proximity as a Resource Base for Competitive Advantage: University-Industry Links for Technology Transfer." *Journal of Technology Transfer* 29(3-4): 311-326.

Link, A. N. 1981a. "Basic Research and Productivity Increase in Manufacturing: Some Additional Evidence." *American Economic Review* 71(5):1111-1112.

Link, A. N. 1981b. *Research and Development Activity in U.S. Manufacturing*. New York: Praeger.

Link, A. N. 1995. *A Generosity of Spirit: The Early History of the Research Triangle Park*. Research Triangle Park, NC: University of North Carolina Press for the Research Triangle Park Foundation.

Link, A. N. 2002. *From Seed to Harvest: The Growth of the Research Triangle Park*. Research Triangle Park, NC: University of North Carolina Press for the Research Triangle Park Foundation.

Link, A. N., and K. R. Link. 2003. "On the Growth of U.S. Science Parks." *Journal of Technology Transfer* 28(1):81-85.

Link, A. N. and J. T. Scott. 1998. *Public Accountability: Evaluating Technology-Based Institutions*. Norwell, MA: Kluwer Academic Publishers.

Link, A. N. and J. T. Scott. 2001. "Public/Private Partnerships: Stimulating Competition in a Dynamic Market." *International Journal of Industrial Organization* 19(5):763-794.

Link, A. N., and J. T. Scott. 2003a. "The Growth of Research Triangle Park." *Small Business Economics* 20(2):167-175.

Link, A. N., and J. T. Scott. 2003b. "U.S. Science Parks: The Diffusion of an Innovation and Its Effects on the Academic Mission of Universities." *International Journal of Industrial Organization* 21(9):1323-1356.

Link, A. N., and J. T. Scott. 2005. "Opening the Ivory Tower's Door: An Analysis of the Determinants of the Formation of U.S. University Spin-Off Companies." *Research Policy* 34(7):1106-1112.

Link, A. N., and J. T. Scott. 2006. "U.S. University Research Parks," *Journal of Productivity Analysis* 25(1):43-55.

Link, A. N., and J. T. Scott. 2007. "The Economics of University Research Parks." *Oxford Review of Economic Policy* 23(4):661-674.

Link, A. N., and D. S. Siegel. 2003. *Technological Change and Economic Performance*. London, UK: Routledge.

Luger, M. I., and H. A. Goldstein. 1991. *Technology in the Garden*. Chapel Hill, NC: University of North Carolina Press.

National Research Council. 1999. *Industry-Laboratory Partnerships: A Review of the Sandia Science and Technology Park Initiative*. Charles W. Wessner, ed. Washington, DC: National Academy Press.

National Research Council. 2001. *A Review of the New Initiatives at the NASA Ames Research Center*. Charles W. Wessner, ed. Washington, DC: National Academy Press.

Nelson, R. R. 1993. *National Innovation Systems: A Comparative Analysis*. New York: Oxford University Press.

Nelson, R. R. and S. G. Winter. 1982. *An Evolutionary Theory of Economic Change*. Cambridge, MA: Harvard University Press.

Phillimore, J. 1999. "Beyond the Linear View of Innovation in Science Park Evaluation: An Analysis of Western Australian Technology Park." *Technovation* 19(11):673-680.

Rothaermel, F. T., and M. C. Thursby. 2005a. "Incubator Firm Failure or Graduation? The Role of University Linkages." *Research Policy* 34(7):1076-1090.

Rothaermel, F. T., and M. C. Thursby. 2005b. "University-Incubator Firm Knowledge Flows: Assessing Their Impact on Incubator Firm Performance." *Research Policy* 34(3):302-320.

Shearmur, R., and D. Doloreux, D. 2000. "Science Parks: Actors or Reactors? Canadian Science Parks in their Urban Context." *Environment and Planning* 32(6):1065-1082.

Siegel, D. S., P. Westhead, and M. Wright. 2003. "Assessing the Impact of Science Parks on Research Productivity: Exploratory Firm-Level Evidence from the United Kingdom." *International Journal of Industrial Organization* 21(9):1357-1369.

Sofouli, E., and N. S. Vonortas. 2007. "S&T Parks and Business Incubators in Middle-Sized Countries: The Case of Greece." *Journal of Technology Transfer* 32(5):525-544.

Sternberg, R. 1990. "The Impact of Innovation Centres on Small Technology-based Firms: The Example of the Federal Republic of Germany." *Small Business Economics* 2(2):105-118.

Swann, G. M. P. 1998. "Towards a Model of Clustering in High-Technology Industries." In G. M. P. Swann, M. Prevezer, and D. Stout, eds. *The Dynamics of Industrial Clustering*. Oxford, UK: Oxford University Press.

Tassey, G. 1997. *The Economics of R&D Policy*. Westport, CT: Quorum Books.

United Nations Educational, Scientific and Cultural Organization (UNESCO). 1994. <http://www.unesco.org/pao/s-parks/what.htm>.

United Kingdom Science Park Association (UKSPA). 2003. <http://www.ukspa.org.uk/default.asp?t=1&channel_id=2374&editorial_id=13661> and <http://www.ukspa.org.uk/?channel_id=2375&editorial_id=13885>.

Vaidyanathan, G. 2008. "Technology Parks in a Developing Country: The Case of India." *Journal of Technology Transfer* 33(3):285-299.

Vedovello, C. 1997. "Science Parks and University-Industry Interaction: Geographical Proximity between the Agents as a Driving Force." *Technovation* 17(9):491-502.

Westhead, P. 1995. "New Owner-Managed Businesses in Rural and Urban Areas in Great Britian: A Matched Pairs Comparison." *Regional Studies* 29(4):367-380.

Westhead, P. 1997. "R&D 'Inputs' and 'Outputs' of Technology-based firms Located On and Off Science Parks." *R&D Management* 27(1):45-61.

Westhead, P., and S. Batstone. 1998. "Independent Technology-based Firms: The Perceived Benefits of a Science Park Location." *Urban Studies* 35(12):2197-2219.

Westhead, P., and M. Cowling. 1995. "Employment Change in Independent Owner-Managed High-Technology Firms in Great Britain." *Small Business Economics* 7(2):111-140.

Westhead, P., and D. Storey. 1994. *An Assessment of Firms Located On and Off Science Parks in the United Kingdom*. London, UK: HMSO.

Westhead, P., and D. Storey. 1997. "Financial Constraints on the Growth of High-Technology Small Firms in the U.K." *Applied Financial Economics* 7(2):197-201.

Westhead, P., D. J. Storey, and M. Cowling. 1995. "An Exploratory Analysis of the Factors Associated with the Survival of Independent High-Technology Firms in Great Britain." In F. Chittenden, M. Robertson, and I. Marshall, eds. *Small Firms: Partnerships for Growth*. London, UK: Paul Chapman.

IV

APPENDIXES

Appendix A

Biographies of Speakers*

M. S. ANANTH

Dr. M. S. Ananth is the director of the Indian Institute of Technology-Madras. He graduated from the AC College of Technology with a gold medal in chemical engineering and obtained his Ph.D. degree in chemical engineering in the area of molecular thermodynamics from the University of Florida in 1972.

He joined IIT-Madras as a faculty in the Department of Chemical Engineering in 1972. He has held various senior positions including head of the department, dean of academic courses, and dean of academic research. Appointed as the director of IIT-Madras in December 2001, he has played a key role in preparing The Strategic Plan of IIT-Madras—Vision 2010.

Dr. Ananth was a visiting professor at Princeton University (1982-1983) and at the University of Colorado (1990-1991). He was a visiting scientist in the National Institute of Standards and Technology in Boulder, Colorado (1990-1991), in RWTH in Aachen, Germany (Summer 1983 and Summer 1999), and a visiting thermodynamics expert at Aspen Tech in Massachusetts (Summer 1991).

Dr. Ananth has taught a whole range of chemical engineering courses and has been consistently rated as a good teacher. He is well known for his deep concern for his students. He was responsible for extensive curricular revisions in the B.Tech. program of IIT-Madras with the introduction of many innovative features. He is also deeply concerned about the quality and reach of engineering education in India. He is the national coordinator for a mega project funded by

*As of March 2008. Appendix includes bios distributed at the symposium.

the MHRD, on Technology Enhanced Learning involving many technical institutions in the country.

Dr. Ananth's research interests are in molecular thermodynamics and mathematical modeling. He has published several papers in international journals and is a referee for many journals in chemical engineering. He has been awarded the Herdillia prize for excellence in basic research in chemical engineering by the Indian Institute of Chemical Engineers and the R.W. Fahien Alumni Award for "Distinguished Professional Contributions" for the year 2003 by the Chemical Engineering Department of the University of Florida. He is a fellow of the Indian Institute of Chemical Engineers and the Indian National Academy of Engineering. He is also a member of the National Manufacturing Competitiveness Council (NMCC) and a member of the Scientific Advisory Committee to the Cabinet. He is a consultant for many chemical industries in South India.

JAMES BARKER

Clemson University's 14th president, James F. Barker, is first and foremost an architect. He came to the office with a clear blueprint of Clemson's future in mind—to become one of the nation's top public universities. To achieve this vision, Mr. Barker led the development of 10-year goals and an action plan built around collaboration, focus, relevance, and academic quality.

The results speak for themselves: Today, Clemson is considered an academic rising star, more than doubling research in just 3 years, launching major new economic development initiatives, dramatically increasing the quality of the student body, and rising to a number 27 ranking among the nation's top public universities. For his leadership, Mr. Barker has been honored with the Buck Mickel Award for Business and Community Leadership by the Greater Greenville Chamber of Commerce, the Anderson Independent-Mail's Pointing the Way Leadership Award, *Greenville Magazine*'s The Cliff's Business Person of the Year Award, and the Order of the Palmetto—the state's highest civilian honor. He also was inducted into the Boys and Girls Clubs Hall of Fame.

Mr. Barker has emerged as a national leader in academic and public service arenas. He currently serves on the NCAA Board of Directors, chairing the Division I Committee, and is past-president of the Commission on Colleges of the Southern Association of Colleges and Schools, the region's accrediting agency, which recently honored him with the James F. Rogers Meritorious Service Award. He was awarded the Executive Leadership Award from the Council for Advancement and Support of Education District III and has been awarded honorary doctoral degrees from S.C. State University and Mars Hill College. He also has served on the Truman Scholars selection committee, was a Cambridge University visiting scholar, and is a member of the Shaw Group Board of Directors, serving on the Audit Committee.

Mr. Barker is a recipient of the National Distinguished Professor Award of the Association of Collegiate Schools of Architecture and served as president of that association. He was named fellow of the American Institute of Architects and has also been a partner in an architectural practice.

As president, Mr. Barker remains committed to the classroom. Each spring, he is part of a team that teaches an undergraduate course exploring "a sense of place" in architecture, literature, and history.

Mr. Barker, a native of Kingsport, Tennessee, earned his bachelor of architecture degree from Clemson in 1970 and his master of architecture and urban design degree from Washington University in St. Louis in 1973.

JEFF BINGAMAN

Jeff Bingaman grew up in Silver City in a family with deep New Mexico small town roots. His father was a science professor at Western New Mexico University, and his mother taught in the public schools. He graduated from high school in Silver City. After graduating from Harvard University, he earned a law degree at Stanford. There he met fellow law student Anne Kovacovich. After graduation, they married and returned to New Mexico, where they both practiced law, and their son, John, was born.

Senator Bingaman was elected New Mexico Attorney General in 1978. In 1982, he won election to the United States Senate, and in 2006, was re-elected to serve a fifth term.

Committees and Responsibilities

- Energy and Natural Resources Committee, *chairman*
- Finance Committee
    - Subcommittee on Energy, Natural Resources, and Infrastructure, *chairman*
    - Subcommittee on International Trade and Global Competitiveness
    - Subcommittee on Health Care
- Health, Education, Labor, and Pensions Committee
    - Subcommittee on Children and Families
    - Subcommittee on Retirement and Aging
- Joint Economic Committee, *senior member*

JANE DAVIES

Jane Davies was appointed chief executive of Manchester Science Park (MSP) in October 2000 and was elected president of the European Division of the International Association of Science Parks (IASP) in 2006. Ms. Davies is also chairman of the UK Science Park Association (UKSPA).

Ms. Davies has a degree in chemistry from St. Anne's College, Oxford, and joined BP Chemicals as a PVC plant chemist on graduation. She spent 18 years with BP in a range of roles including international oil trader in New York and regional manager of BP's international aviation business, Air BP. During this period she also spent two years on secondment to the FCO Planning Staff. On leaving BP, Jane ran the Buxton Festival for four years, where she gained valuable experience in running a micro business with insufficient resources—circumstances that many of her tenants have to deal with.

In Ms. Davies' time at MSP, the company has expanded its operations to facilities on four sites in Manchester and has invested in people and processes to strengthen the links between its tenant companies and its university shareholders. The science park and its tenants are seen as an important part of Manchester's knowledge economy.

PETER ENGARDIO

Pete Engardio is a senior writer for *BusinessWeek*, focusing on global business and economic trends.

Mr. Engardio joined *BusinessWeek* in 1985 as a correspondent in the magazine's Atlanta bureau. In 1987 he moved to Miami as bureau manager. In 1990 he became a correspondent in the Hong Kong bureau, where he covered Asian business for six years. In 1996 he moved to New York as an Asian editor. From 1998 to 2001, he was editor of the Asian Edition.

In 2005, Mr. Engardio anchored *BusinessWeek*'s special issue "China & India: What You Need to Know," winner of the Institute for Political Journalism Award. His 2004 cover "The China Price" won an Overseas Press Club (OPC) Award. In 2003, he anchored two groundbreaking covers on offshore outsourcing of skilled work, "Is Your Job Next" and the "Rise of India," for which he received George Polk, Loeb, and Sigma Delta Chi awards, and was named a finalist for a National Magazine Award. The pieces sparked congressional hearings and a national debate on outsourcing. He also won a Harry Chapin award sponsored by World Hunger Year for his 2002 cover "Fighting Poverty" and a Clarion Award and OPC citation in 2001 for his cover "Global Capitalism: Can it be Made to Work Better?" In 1996, he received an OPC Award for his international cover story, "China's New Elite." In 1997 he received an OPC citation for "Asia: Time for a Reality Check," written as he was finishing his Hong Kong tour, and he was part of the *BusinessWeek* Asia Team that won a 1998 Overseas Press Club Award for coverage of Asia in Crisis.

Prior to joining *BusinessWeek*, Mr. Engardio was a feature editor for *Business Korea* in Seoul, as well as a stringer for *BusinessWeek*. Before that, he worked for the Bay City News Service in San Francisco and the *San Francisco Bay Guardian*.

Mr. Engardio edited and wrote opening chapters for the book *Chindia: How China and India are Revolutionizing Global Business*, published in 2007 by McGraw-Hill, and is co-author with Mark L. Clifford of *Meltdown: Asia's Boom, Bust, and Beyond*, published in 2000 by Prentice-Hall. Mr. Engardio's recent speaking appearances include McGill University and HEC business school (Montreal), the Conference Board (New York), National Public Radio, CNN International, CNBC, the Center for Global Development (Washington, DC), and Duke University.

In 2004, Mr. Engardio was a Reuters Journalism Fellow at Oxford University. He holds a B.A. from Central Michigan University and an M.A. from the University of Missouri School of Journalism.

CHRISTINA GABRIEL

Christina Gabriel joined The Heinz Endowments in 2006 with extensive experience in research management, university-industry collaboration, and technology transfer. She is responsible for the foundation's efforts to capitalize on the research strengths of the region's universities, medical centers, corporate and government laboratories to promote economic growth and opportunity in southwestern Pennsylvania.

After receiving her doctorate in electrical engineering and computer science from the Massachusetts Institute of Technology (MIT), Dr. Gabriel began her professional career conducting experimental research at AT&T Bell Laboratories in New Jersey. Her work focused on lasers, optical fibers and thin-film waveguide devices for telecommunications, switching and computing applications. She holds three patents.

Dr. Gabriel later joined the National Science Foundation, first to direct industry-university collaborative centers programs and later as deputy head of the $350-million engineering directorate. While in Washington, DC, she spent about a year on Capitol Hill as a professional staff member for the U.S. Senate Committee on Appropriations. She also has been a visiting professor at the University of Tokyo in Japan. From 1998 to 2006, she worked at Carnegie Mellon University in Pittsburgh, eventually becoming vice provost and chief technology officer.

Dr. Gabriel received both her master's and doctoral degrees from MIT and her undergraduate degree in electrical engineering from the University of Pittsburgh. She has been a review panelist and advisor for the National Science Foundation and the National Academies and serves on the MIT Corporation Visiting Committee on Sponsored Research. In Pittsburgh, she has served on several nonprofit boards and as an external technology advisor for the Pittsburgh Public Schools' strategic planning process. She was an AT&T Bell Laboratories GRPW Fellow and a national merit scholar.

MARY GOOD

Mary L. Good is the Donaghey University Professor at the University of Arkansas at Little Rock and serves as dean for the College of Information Science and Systems Engineering. She is managing member for the Fund for Arkansas' Future, LLC (an investment fund for startup and early-stage companies), past president of the American Association for the Advancement of Science, past president of the ACS, and an elected member of the National Academy of Engineering. She presently serves on the Boards of Acxiom, Inc., St. Vincent Health System, and Delta Bank and Trust.

Previously she served a four-year term as the Under Secretary for Technology for the Technology Administration in the Department of Commerce, a presidentially appointed, Senate-confirmed position. In addition, she chaired the National Science and Technology Council's Committee on Technological Innovation (NSTC/CTI), and served on the NSTC Committee on National Security. Previously she has served as the senior vice president for technology for Allied Signal and as the Boyd Professor of Chemistry and Materials Science at Louisiana State University.

Dr. Good was appointed to the National Science Board by President Carter in 1980 and by President Reagan in 1986. She was the chair of that board from 1988 to 1991, when she received an appointment by President Bush to be a member of the President's Council of Advisors on Science and Technology.

Dr. Good has received many awards, including the National Science Foundation's Distinguished Public Service Award, the American Institute of Chemists' Gold Medal, the Priestly Medal from the American Chemical Society, and the Vannevar Bush Award from the National Science Board, among others.

Dr. Good received her bachelor's degree in chemistry from the University of Central Arkansas and her M.S. and Ph.D. degrees in inorganic chemistry from the University of Arkansas at Fayetteville.

DAVID HOLDEN

David Holden is responsible for strategic marketing for the Minatec center, located in Grenoble, France. As a member of the core management team, Dr. Holden actively identifies and engages international partners as part of the strategic development plan for the center. In particular, his focus is on commercialization and technology transfer with private enterprise in the areas of nanotechnology, microsystems, renewable energy, and new technologies for healthcare.

Prior to joining CEA in 2003, Dr. Holden provided consulting services to commercial microsystems providers in the United States and Europe on technologies for inertial sensors, biochips, and inkjet printing technology. From 1999 to 2001, he was responsible for polymer microfluidics development for inkjet products at Xerox Research, with several successful technology platforms transferred to commercial production. From 1993 to 1999, Dr. Holden was involved with various startup companies in flat-panel display technologies, including Pixtech, AVT, and ADT, a Robert Bosch GmbH development stage subsidiary based in Stuttgart. Dr. Holden began his professional career in 1986 as a process engineer at General Electric corporate R&D where he developed TFT LCD technology that he later transferred to Thomson in France.

Dr. Holden received his MBA from INSEAD (1993) and a B.S. in materials science and engineering from Cornell University (1986).

As a representative of the MINATEC center, Dr. Holden frequently represents the activities of the greater Grenoble area, highlighting links between research, education and industry to stimulate economic development in the high-tech sector.

VICTOR LECHTENBERG

Victor L. Lechtenberg was named Purdue University's interim provost in July 2007. He succeeds Sally Mason, who left Purdue to become president of the University of Iowa.

Since 2004, Dr. Lechtenberg had served as Purdue's vice provost for engagement, working to align the university's intellectual and other resources to assist in Indiana's economic growth and to address challenges facing the state's business and industry. Prior to that, he served for 10 years as Purdue's dean of agriculture.

As interim provost, Dr. Lechtenberg is responsible for oversight of campus management in research, engagement, teaching, and related academic activities in coordination with the Office of the President. His office also oversees academic systems such as the libraries, computing center, and student services.

Dr. Lechtenberg has been an active leader on both state and national levels with respect to research and technology policy. He has also been an advocate for technology-related economic growth in the food, agriculture, and natural resource sectors. He chaired the USDA's National Agricultural Research, Extension, Education, and Economics Advisory Board from 1996 to 2002.

In addition, Dr. Lechtenberg has served as president of the Crop Science Society of America and as president of the Council for Agricultural Science and Technology (CAST). He is a fellow of the American Society of Agronomy and Crop Science Society of America. He also recently served on the National Academies Division of Earth and Life Studies Committee and currently serves on the National Academies Board on Agriculture and Natural Resources.

Dr. Lechtenberg received his B.S. degree in 1967 from the University of Nebraska, where he was an agriculture honors program graduate. He received his Ph.D. in agronomy from Purdue University in 1971.

YENA LIM

Ms. Yena Lim joined the Agency for Science, Technology and Research (A*STAR) in June 2006. She assumed the position of managing director A*STAR in January 2007. A*STAR has 232 staff in corporate headquarters, 3,267 scientists and researchers and related research support staff at the research institutes, and 69 staff focusing on R&D commercialization. The organization strives to foster world-class scientific research and talent for a vibrant knowledge-based economy by strengthening the stock of human, intellectual, and industrial capital. It has a budget of S$5.4 billion over 5 years to undertake R&D activities and to carry out its strategies.

A*STAR is Singapore's lead agency for fostering world-class scientific research and talent for a vibrant knowledge-based Singapore. A*STAR actively nurtures public-sector research and development in biomedical sciences, physical sciences, and engineering, with a particular focus on fields essential to Singapore's manufacturing industry and new growth industries. It oversees 14 research institutes and supports extramural research with the universities, hospital research centers, and other local and international partners. At the heart of this knowledge-intensive work is human capital. Top local and international scientific talent drive knowledge creation at A*STAR research institutes. The agency also sends scholars for undergraduate, graduate, and post-doctoral training in the best universities, a reflection of the high priority A*STAR places on nurturing the next generation of scientific talent.

Before A*STAR, Ms. Lim was with the Ministry of Trade and Industry (MTI), as the director of its Resource Division. She oversaw economic policies pertaining to competitiveness and capability development for the economy. This included industrial land use planning and policies, energy security and reforms in the electricity and gas markets, support of manpower development and research and development strategies to support economic growth. Her team at MTI came up with the Science and Technology Plan 2010 in close collaboration with A*STAR and other agencies. They also provided staff support to the Ministerial Committee on Research and Development that resulted in the creation of the Research, Innovation and Enterprise Council chaired by PM, and the National Research Foundation.

As an officer of the Administrative Service, Ms. Lim has also worked in the Ministry of Transport overseeing the land transport portfolio; the Public Service Division of the Prime Minister's Office looking after human resources policies for the civil service; the Ministry of Education overseeing the school physical infrastructure planning and development portfolio and higher education policy; and the Ministry of Finance covering the Government's fiscal policy, including revenue and taxation policies.

Ms. Lim has served on the boards of Sentosa Development Corporation, Jurong Port, Nanyang Polytechnic, the Singapore-MIT alliance, the Duke-NUS

Graduate Medical School, and the Workforce Development Agency, and she is currently on the board of A*STAR and a panel member of the Enterprise Challenge.

ALBERT LINK

Albert N. Link is professor of economics at the University of North Carolina at Greensboro. He received a B.S. degree in mathematics from the University of Richmond and a Ph.D. degree in economics from Tulane University. His research focuses on innovation policy, university entrepreneurship, and the economics of R&D. He is the editor-in-chief of the *Journal of Technology Transfer*. Professor Link's most recent books include: *The Economic Theory of Invention and Innovation* (Edward Elgar, 2008), *Cyber Security: Economic Strategies and Public Policy Alternatives* (Edward Elgar, 2008), *Entrepreneurship, Innovation, and Technological Change* (Oxford University Press, 2007), *Public/Private Partnerships* (Springer, 2006), and *Evaluating Public Research Institutions* (Springer, 2005). In addition, he is the author of the two-volume history of Research Triangle Park in North Carolina: *A Generosity of Spirit: The Early History of the Research Triangle Park* (1995) and *From Seed to Harvest: The Growth of the Research Triangle Park* (2002), both published by the University of North Carolina Press for the Research Triangle Foundation of North Carolina. Much of Professor Link's research has been funded by the National Science Foundation, the OECD, the World Bank, and various science and technology ministries in developed nations. Currently, Professor Link is serving as the vice-chairperson of the Innovation and Competitiveness Policies Committee of the United Nation's Economic Commission for Europe (UNECE).

ROBERT MCMAHAN

Robert McMahan is senior advisor to the Governor of North Carolina for science and technology and the executive director of the North Carolina Board of Science and Technology. In this role he advises the Governor, the Secretary of Commerce, the General Assembly, and the Boards of Science and Technology and Economic Development about science and technology matters and supports and advises the state government on science, technology, entrepreneurship and technology-based economic development. He also serves as the primary liaison to the University of North Carolina System, the SBTDC, the NC Community College System, other private colleges and universities, key agencies such as the Biotechnology Center and MCNC, and associations such as CED, NCTA, and NCBIO with regard to these issues.

Prior to this Dr. McMahan was a senior technology strategist for In-Q-Tel, a private venture capital organization funded by the CIA and NIMA, where he was part of a team responsible for developing a technology investment strategy for the CIA, and then deriving, molding, and structuring individual investments and technologies within the portfolio. Before joining In-Q-Tel, he was executive vice president of engineering/research and development for the Swiss-based, mid-cap GretagMacbeth, LLC, where he was responsible for the company's world-wide research, engineering, and product development activities and for the creation and operation of the company's Advanced Technology Laboratories in the Research Triangle Park. He joined GretagMacbeth after its acquisition in 2000 of McMahan Research Laboratories, the advanced technologies company for which he was president and CEO and which he founded in 1987 in Cambridge, MA, and expanded to the RTP in 1989. Dr. McMahan has been involved in the creation of a number of technology startups, and he has participated in equity and LBO capital raises.

In addition to his duties with the state, Dr. McMahan also currently holds the positions of research professor of physics and astronomy at the University of North Carolina at Chapel Hill, where he teaches and conducts research in cosmology, instrumentation, and the large-scale structure of the universe, and adjunct professor of technology and management at the North Carolina State University College of Textiles. Dr. McMahan received bachelor's degrees in physics and in the history of art from Duke University in 1982, a Ph.D. degree in physics from Dartmouth in 1986, and a postdoctoral fellowship from the Harvard University/Smithsonian Astrophysical Observatory Center for Astrophysics. He has published over forty papers in scientific and engineering journals, sits on a number of state and corporate boards and commissions, and holds multiple patents in the field.

C. D. MOTE, JR.

Currently the president of the University of Maryland, C. D. "Dan" Mote, Jr., spent most of his life in Berkeley. He attended two grammar schools there, earned three degrees in mechanical engineering at the University of California, Berkeley, and spent 31 years on the faculty. He was chair of the mechanical engineering department, held an endowed chair in mechanical systems and then became vice chancellor for university relations in order to create and lead Berkeley's comprehensive capital campaign to raise $1.1 billion by 2001.

Dr. Mote's research lies in dynamic systems and biomechanics. Internationally recognized for work on the dynamics of high-speed rotating and translating materials and the biomechanics of snow skiing injuries, he has authored and co-authored more than 300 publications, holds patents in the United States, Norway, Finland, and Sweden, and mentored 56 Ph.D. students while at the University of California.

In 1998 Dr. Mote was appointed president of the University of Maryland and Glenn L. Martin Institute Professor of Engineering. He was recruited to lead the University of Maryland to national eminence under a mandate by the state.

He is now leading a $1-billion capital campaign so that Maryland can maintain affordable access for state residents.

His goals call for the university to lead the state in the development of its high-tech economy, especially in information, bioscience and biotechnology, energy, language, security, and nanotechnology. He has greatly expanded the university's partnerships with federal laboratories and inaugurated the first research park sponsored by the People's Republic of China outside the mainland. China also founded its first international Chinese language, literature, and culture center at Maryland. Under his leadership, the university founded a research park on 128 acres adjacent to the campus with 3 million square feet of development potential, making it the largest park in Maryland and Greater Washington. The NOAA National Center for Weather and Climate Prediction will be located there.

President Mote received the Berkeley Distinguished Teaching Award and the Berkeley Citation, and he was named a distinguished engineering alumnus. He has received three honorary doctorates, is a member and councilor of the U.S. National Academy of Engineering, and has received its prestigious Founder's Award. He was elected to honorary membership in the American Society of Mechanical Engineers International, and he received the Humboldt Prize of the Federal Republic of Germany. He is a fellow of the American Academy of Arts and Sciences, the Acoustical Society of America, and the American Association for the Advancement of Science.

JOHN NIEDERHUBER

John E. Niederhuber, M.D., became director of the National Cancer Institute (NCI) in September 2006. Prior to that he had been the Institute's acting director from June 2006. He was formerly a professor of surgery and oncology at the University of Wisconsin School of Medicine. Dr. Niederhuber served the University of Wisconsin as the director of the University of Wisconsin Comprehensive Cancer Center from July 1997 until October 2002. He came to the University of Wisconsin in 1997 from Stanford University where he had served as chair of the Department of Surgery. In June 2002, President George W. Bush appointed Dr. Niederhuber chair of the National Cancer Advisory Board, a position he held until resigning to become the deputy director at NCI in 2005.

Dr. Niederhuber's research at NCI focuses on the study of tissue stem cells as the cell-of-origin for cancer. His lab is working to identify, characterize fully, and isolate this population of cells with the hypothesis that such cells might be the required therapeutic target. His lab is also studying the viral cancer vector HPV, to identify the binding site theorized to be a stem cell epithelial receptor.

Dr. Niederhuber is a nationally recognized cancer surgeon with a special clinical emphasis in gastrointestinal cancer, hepatobiliary cancer, and breast cancer. He is recognized for his pioneering work in hepatic artery infusion chemotherapy and was the first to demonstrate the feasibility of totally implantable vascular access devices.

Dr. Niederhuber is a graduate of Bethany College, Bethany, West Virginia, and the Ohio State University School of Medicine. He was an NIH academic trainee in Surgery at the University of Michigan (1969-1970) and a visiting fellow, Division of Immunology, The Karolinska Institute, Stockholm, Sweden (1970-1971). He completed his training in surgery at the University of Michigan in 1973. He was a member of the faculty of the University of Michigan from 1973 to 1987, being promoted to professor of microbiology/immunology and professor of surgery in 1980.

JAIME PARADA

Appointed in September 2007, Mr. Parada is the president and CEO of the Program Monterrey City of Knowledge and of the Institute of Innovation and Technology Transfer of Nuevo Leon, responsible for the Research and Innovation Technology Park. From December 2000 to 2005, Mr. Parada was the director general of the National Council on Science and Technology (CONACYT), Mexico's equivalent to a Ministry of Science and Technology. CONACYT is the cabinet-level agency directly answerable to the president, responsible for science and technology policy in Mexico and entrusted with the task of funding R&D activities through several institutional funds, graduate scholarships for Mexican students, and the National Researcher's System, whereby top researchers get supplemental funding to support their research activities. CONACYT also coordinates 27 research centers in all fields of knowledge, as well as the National Science and Technology Registry and Mexico's tax incentive scheme for companies involved in R&D. Mr. Parada graduated as a mechanical-electrical engineer from UNAM's College of Engineering. He took postgraduate and specialization courses in both Mexican and foreign institutions in the areas of strategic planning, quality assurance systems, manufacturing systems, research and technological development, technology management, and teaching and teaching methods. He has 25 years of experience in the fields of Technological R&D in the government, academy, research areas, and private enterprises.

PHILLIP H. PHAN

Phillip H. Phan is Warren H. Bruggeman '46 and Pauline Urban Bruggeman Distinguished Professor of Management at the Lally School of Management & Technology at Rensselaer Polytechnic Institute. He was the 2004/2005 Haniel Foundation Visiting Professor at Humboldt University in Berlin and is the 2006/2007 Robert Bosch Foundation Public Policy Fellow at the American Academy in Berlin. He is the summer 2007/2008 visiting research professor at Singapore Management University.

In 2007/2008 Dr. Phan published three books including *Theoretical Advances in Family Enterprise Research* (InfoAge Press), *Entrepreneurship and Economic Development in Emerging Regions* (Edward Elgar), and *Taking Back the Boardroom* (2nd ed) (Imperial College Press). His areas of research are in innovation and technology transfer, corporate governance in strategic alliances, and entrepreneurial market-entry strategies. Dr. Phan has published more than 80 research articles in such journals as the *Academy of Management Journal*, *Asia Pacific Journal of Management*, *Corporate Governance*, *European Management Journal*, *Journal of Business Venturing*, *Journal of International Business Studies*, *Journal of Management*, *Research Policy*, *IEEE Transactions Engineering Management*, and *Small Business Economics*. He served on the editorial review board of the *Academy of Management Journal* and is now associate editor for the *Journal of Business Venturing*, *Journal of Financial Stability*, and *Journal of Technology Transfer*.

Dr. Phan has consulted for the World Bank, OECD, Singapore Institute of Directors, U.S. Small Business Administration, and such companies as Motorola, HP, IBM, Ernst & Young, Metsaliitto (Finland), Pillsbury, Finlombardia (Italy), Agilent Technologies, SK Group (Korea), SanomaWSOY (Finland), Singapore Airlines, PACCAR (United States), and technology venture capital firms in Toronto and New York.

Dr. Phan has been a regular expert contributor to CNBC, Bloomberg News, CNNOnline, the *Wall Street Journal,* and various regional and national print and media outlets in Singapore, Germany, and Canada.

LAWRENCE SCHUETTE

Dr. Schuette entered the Senior Executive Service in July 2007 and is currently the senior civilian responsible for innovation at the Office of Naval Research. This portfolio includes the Innovative Naval Prototypes and Swampworks projects. These are high-risk/high-payoff technology investments that are potentially "game changing" or "disruptive" in nature.

From September 2006 to July 2007, he served as a special assistant to the Assistant Secretary of the Navy for Research Development and Acquisition. As a special assistant, Dr. Schuette focused on in-house RDT&E workforce issues.

Dr. Schuette served as the deputy chairman of the Joint IED Defeat Organization (JIEDDO) Laboratory Board from March 2006 to September 2006. In this capacity, Dr. Schuette created the science and technology strategy used by JIEDDO to combat improvised explosive devices.

Dr. Schuette headed the Innovative Systems Subgroup of the OSD Technical Joint Cross Service Group during Base Realignment and Closure 2005 from August 2003 to August 2005. Dr. Schuette was responsible for assessing and recommending for potential consolidation and closure of the research activities of the Department of Defense.

He started his 20 years of civilian service in 1987 as a research scientist in the Acoustics Division of the Naval Research Laboratory working on advanced signal processing and computer visualization.

Dr. Schuette attended The Catholic University of America and graduated in 1983 with a bachelor's degree in electrical engineering. Following graduation, he worked on underwater fiber optic sensors as a contractor at the Naval Research Laboratory in the Acoustics Division. Dr. Schuette completed his master's degree in electrical engineering at The Catholic University of America in 1985. In 1995, Dr. Schuette received his Ph.D. from The Catholic University of America in electrical engineering.

Dr. Schuette's awards include the Secretary of Defense's award for Exceptional Civilian Service, the Naval Meritorious Unit Commendation, and the Naval Unit Commendation. He is a member of the National Defense Industry Association and the United States Naval Institute.

ZHU SHEN

Zhu Shen is the CEO of BioForesight, a life science strategic consulting company providing licensing, partnering, financing, product launch, and public relations services to clients in the United States and China. She was vice president of business development at ItherX Pharmaceuticals (formerly Immusol), a private biopharmaceutical company in San Diego developing novel therapeutics in oncology and HCV. Her responsibilities include leading ItherX's licensing, partnering, and corporate communication. She also held positions at The Wilkerson Group/IBM, Bayer, and Chiron, with increasing responsibilities in marketing, strategic planning, and business development.

Dr. Shen is an organizer, speaker, and chair of numerous life science business conferences on licensing, venture investing, and cross-Pacific partnering/CRO trends. She is vice chairwoman and serves on the board of directors of Sino-American Biomedical & Pharmaceutical Professionals Association (SABPA), as well as the BIOCOM Asia Task Force Steering Committee, the Athena San Diego Bioscience Committee and CFO Committee. Dr. Shen has authored numerous articles published in *Pharmaceutical Executive, Ernst & Young Global Pharmaceutical Reports,* and *BioExecutive International.* She has been interviewed by the *Wall Street Journal*, *BusinessWeek*, *Los Angeles Times*, *China Economic Review*, *Drug Discovery News*, *Genetic Engineering News*, *San Diego Union Tribune*, *San Diego Business Journa*l, and *ASIA Media*.

Dr. Shen received the 2006 Annual Asian Heritage Award in Science, Technology, and Research, and the SABPA Achievement Award in 2006 and 2007. She received her Ph.D. in biochemistry from the University of Colorado and an MBA from the Johnson School at Cornell University. She attended Peking University and Peking Union Medical College in China.

RICHARD STULEN

Richard Stulen is the chief technical officer and vice president of Science and Technology & Research Foundations at Sandia National Laboratories. In his line role, he is responsible for R&D activities of over 1,300 scientists and engineers working in nanoscience and technology, materials science, advanced fusion and pulsed power technology, high-performance computing, radiation sciences, microelectronics and microsystems, and engineering sciences. Dr. Stulen is also responsible for Sandia's technology transfer program, the Science, Technology, and Engineering Strategic Management Unit, and the Nuclear Weapons Program's Science and Technology programs.

Dr. Stulen earned his Ph.D. degree in solid state physics from Purdue University and joined Sandia National Laboratories as a member of the Technical Staff in 1976. During his career, he has organized and chaired international workshops and published extensively in areas related to surface science and Extreme Ultraviolet (EUV) lithography. In 1999, he received Lockheed Martin's prestigious NOVA award for technical excellence. His previous positions include director of the Exploratory Systems and Development Center focusing on homeland security and the CEO and COO of the Extreme Ultraviolet Lithography Virtual National Laboratory. He served on the 2003 DoD Defense Science Board Summer Study on Homeland Security and is currently on the board for the New Mexico Center for Advanced Computing.

JAMES TURNER

Jim Turner, chief counsel of the Committee on Science and Technology, has 30 years of experience as a congressional staff member working on technology and energy policy. He graduated from Georgetown, Yale, and Westminster College. He was a Clinton Presidential Transition Team member for the Department of Commerce.

Mr. Turner is a trustee of the University of Virginia's engineering school (UVA/SEAS) and a Dean's Advisory Council member for Carnegie Mellon's Heinz School. He serves on the board of directors of Scientists and Engineers for America and the American National Standards Institute (ANSI). He chairs UVA/SEAS's Advisory Board for the Science, Technology, and Society program. With Bill Bonvillian, he does Washington coordination for the joint MIT/UVA Washington Summer Internship program.

Mr. Turner has received standards medals from ASME, ANSI, and ASTM, and awards from the Virginia Engineering Foundation, the Federal Patent Lawyer Association, the Technology Transfer Society, the National Institute of Building Sciences, the Federal Laboratory Consortium, the Semiconductor Industry Association, and the U.S. Metric Association.

ILONA VASS

Ilona Vass graduated in chemical system engineering at the Chemical University of Veszprém, Hungary. She received her doctorate in applied mathematics. She started her career at the Research Institute for Technical Chemistry. She was responsible for the mathematical modelling of chemical processes. At the Technical University of Budapest her responsibility was research and lecturing in chemometrics, artificial intelligence, and expert and quality assurance systems. Richter Gedeon Pharmaceutical Works provided an opportunity to work on the realization of Laboratory Information System. In 1992 she became the deputy director general of the Department of Innovation at the Ministry of Industry and Trade. From 1996 she worked as the deputy head of EUREKA Secretariat, Brussels. In 2000 she became the director of the R&D Unit, Ericsson Telecommunicaton Hungary Ltd. In 2004 she was appointed to be the general vice president of the National Office for Research and Technology, Hungary.

RICK L. WEDDLE

Rick Weddle is president and CEO of the Research Triangle Foundation of North Carolina, owner and developer of the renowned Research Triangle Park (RTP). Mr. Weddle oversees park operations and development and is setting the strategic direction for RTP's future. Under his leadership to date, the park's success has surpassed its historic performance. Since 2004, RTP has generated successful development projects with projected capital investment of over $800 million and the projected creation of over 6,300 new, high-quality jobs.

Mr. Weddle's career encompasses over 25 years of successful leadership and organizational change management. Previously, he led regional economic development organizations in four different states, including the Greater Phoenix Economic Council; the Toledo, Ohio, Regional Growth Partnership; the San Joaquin Partnership & Business Council in Stockton, California; and Winston-Salem Business, Inc., in North Carolina. During his tenure, these organizations created a combined total of 32,000 new jobs and invested over $3.4 billion in their regions.

An active speaker and leader in numerous economic development and science park organizations, Mr. Weddle was elected president of the International Association of Science Parks (IASP) North American Division in December 2007. During his two-year term as N.A. Division President, Mr. Weddle will serve on IASP's International Board of Directors, and the Research Triangle Park will host the 2009 IASP World Conference on Science and Technology Parks. Active in the International Economic Development Council (IEDC) since its inception, Mr. Weddle, a fellow member, was elected as the first chairman of the board of IEDC in 2002 and received the designation of honorary life member in 2007. Currently, Mr. Weddle serves on the board of directors of the Association of

University Research Parks (AURP), Research Triangle Regional Partnership; vice chair of Government Affairs, Legislation and Policy for the Regional Transportation Alliance; and co-chair of Reality Check, a joint effort of the Urban Land Institute and Triangle Tomorrow.

CHARLES WESSNER

Charles Wessner is a National Academy scholar and director of the Program on Technology, Innovation, and Entrepreneurship. He is recognized nationally and internationally for his expertise on innovation policy, including public-private partnerships, entrepreneurship, early-stage financing for new firms, and the special needs and benefits of high-technology industry. He testifies to the U.S. Congress and major national commissions, advises agencies of the U.S. government and international organizations, and lectures at major universities in the United States and abroad. Reflecting the strong global interest in innovation, he is frequently asked to address issues of shared policy interest with foreign governments, universities, and research institutes, often briefing government ministers and senior officials from around the world.

Dr. Wessner's work addresses the linkages between science-based economic growth, entrepreneurship, new technology development, university-industry clusters, regional development, small-firm finance, and public-private partnerships. His program at the National Academies also addresses policy issues associated with international technology cooperation, investment, and trade in high-technology industries. Currently, he directs a series of studies centered on government measures to encourage entrepreneurship and support the development of new technologies. Foremost among these is a congressionally mandated study of the Small Business Innovation Research (SBIR) Program, reviewing the operation and achievements of this $2 billion award program for small companies and startups. Under the auspices of the Board on Science, Technology, and Economic Policy, he is also directing a major study on best practice in global innovation programs, entitled *Comparative Innovation Policy: Best Practice for the 21st Century*, and a complementary analysis entitled *Competing in the 21st Century: Best Practice in State & Regional Innovation Initiatives*.

S. PETE WORDEN

S. Pete Worden (Brig. Gen., USAF, ret.) is the NASA Ames Research Center director. Prior to becoming director, Dr. Worden was a research professor of astronomy, optical sciences, and planetary sciences at the University of Arizona where his primary research direction was the development of large space optics for national security and scientific purposes and near-Earth asteroids. Additionally he worked on topics related to space exploration and solar-type activity in nearby stars. He is a recognized expert on space issues—both civil and military.

Dr. Worden has authored or co-authored more than 150 scientific technical papers in astrophysics, space sciences, and strategic studies. Moreover, he served as a scientific co-investigator for two NASA space science missions.

In addition to his former position with the University of Arizona, Dr. Worden served as a consultant to the Defense Advanced Research Projects Agency (DARPA) on space-related issues. During the 2004 congressional session Dr. Worden worked as a congressional fellow with the Office of Senator Sam Brownback (R-KS), where he served as Senator Brownback's chief advisor on NASA and space issues.

Dr. Worden retired in 2004 after 29 years of active service in the United States Air Force. His final position was director of development and transformation, Space and Missile Systems Center, Air Force Space Command, Los Angeles Air Force Base, California. In this position he was responsible for developing new directions for Air Force Space Command programs and was instrumental in initiating a major Responsive Space Program designed to produce space systems and launchers capable of tailored military effects on timescales of hours.

Dr. Worden was commissioned in 1971 after receiving a B.S. degree from the University of Michigan. He entered the Air Force in 1975 after graduating from the University of Arizona with a doctorate in astronomy. Throughout the 1980s and early 1990s, Dr. Worden served in every phase of development, international negotiations, and implementation of the Strategic Defense Initiative, a primary component in ending the Cold War. He twice served in the Executive Office of the President. As the staff officer for initiatives in the George Bush administration's National Space Council, Dr. Worden spearheaded efforts to revitalize U.S. civil space exploration and earth monitoring programs.

Dr. Worden commanded the 50th Space Wing that is responsible for more than 60 Department of Defense satellites and more than 6,000 people at 23 worldwide locations. He then served as deputy director for requirements at headquarters Air Force Space Command, as well as the deputy director for command and control with the Office of the Deputy Chief of Staff for Air and Space Operations at Air Force headquarters. Prior to assuming his current position, Dr. Worden was responsible for policy and direction of five mission areas: force enhancement, space support, space control, force application, and computer network defense.

Dr. Worden has written or co-written more than 150 scientific technical papers in astrophysics, space sciences, and strategic studies. He was a scientific co-investigator for two NASA space science missions.

Appendix B

Participants List*

Robert Adams
Lake Nona Property Holdings, LLC

Khalid Al-Gahtani
Riyahd Techno Valley

Mazyad Alterkawi
Riyahd Techno Valley

Elaine Amir
Johns Hopkins Montgomery County

M. S. Ananth
Indian Institute of Technology-Madras

Eric Anderson
The Facility Group

Larry Arthur
SAIC-Frederick, Inc.

Ahmed Bakarman
Riyahd Techno Valley

James Barker
Clemson University

William Barnett
Booz Allen Hamilton

Austin Beggs
Innovation Place

Bill Behn
Department of State

Mark Betteridge
Discovery Parks Trust

Hon. Jeff Bingaman
United States Senate

Missy Blair
Association of University Research Parks

Laurent Bochereau
Delegation of the European Commission to the United States

*Speakers in italics.

Michael Bowman
Association of University Research Parks

Kurt Brost
Sanford Research Park Development

Dave Bufter
SAIC-Frederick, Inc.

Laura Butcher
Hershey Center for Applied Research

Lynda Carlson
National Science Foundation

Katie Cash
The Facility Group

Cherrie Chalifoux
CUH2A Architecture, Engineering Planning

Kathryn Clay
U.S. Senate Energy Committee

Jim Clinch
Sandia Science and Technology Park

Mark Crowell
University of North Carolina at Greensboro

Jim Currie
Ohio Agricultural Research & Development Center

Frank D'Alonzo
Fitness and Wellness Professional Services

Brian Darmody
University of Maryland

Jane Davies
Manchester Science Park, United Kingdom

Rick Davis
Chamber of Commerce of Huntsville/Madison County

Gregory Deason
Purdue University Foundation

Casey Deraad
AFRL RVOT

David Dierksheide
National Research Council

Charles Dilks
Dilks Consulting Inc.

Michael Donovan
Boston University

Peter Engardio
BusinessWeek

Julie Evans
UMB BioPark

Gayle Farris
Forest City Boston

Andrew Favata
Core Communities

Terry Foegler
Science and Technology Campus Corporation (SciTech)

John Fonner
BHDP Architecture

Alton Fryer
Mannekin Company

Christina Gabriel
The Heinz Endowments

Camille Gagnon
Innovitech Inc.

Dale Gann
Vancouver Island Technology Park

Luiz Gargione
University of Vale do Paraiba Tech Park

Gino Gemignani
The Whiting-Turner Contracting Company

Robert Geolas
Clemson University International Center for Automotive Research

Tim George
George, Henry, George Partners

Lisa Goble
University of North Carolina at Greensboro

Mary Good
University of Arkansas at Little Rock

Jo Anne Goodnight
National Institutes of Health

Norma Grace
University of New Orleans

Tony Grindberg
NDSU Research & Technology Park

Thomas Hall, III
University Financing Foundation, Inc.

Stephen Hanssen
Wexford Science & Technology, LLC

John Hardin
North Carolina Department of Commerce

Lucas Hargraves
Little Rock Regional Chamber of Commerce

George Harker
University of New Orleans

Tim Harris
National Cancer Institute

Chris Hayter
National Governors Association

David Holden
Minatec

Kevin Holsapple
Los Alamos Research Park

D'Anne Hurd
Worcester Polytechnic Institute

Gregory Hyer
University Research Park University of Wisconsin-Madison

Michel Israël
Embassy of France

Christian Joergens
Embassy of Germany

William Kittredge
U.S. Department of Commerce

Andrew Klein
National Institute of Standards & Technology

Nancy Kossan
University of California, San Diego

David Landry
Sanford Health

Jorge Las Heras
Fundacion Valle Lo Aguirre Universidad de Chile

Mary Pat Lawrence
Sonnenschein Nath & Rosenthal LLP

Victor Lechtenberg
Purdue University

Dan Leri
Innovation Park at Penn State

Scott Levitan
Forest City Science & Technology Group

Yena Lim
Singapore Agency for Science, Technology and Research

Albert Link
University of North Carolina at Greensboro

Jamie Link
Science and Technology Policy Institute

Dennis Lyndon
D.C. Lyndon & Associates, Ltd.

Neil MacDonald
Federal Technology Watch

Joel Marcus
Alexandria Real Estate Equities Inc.

Fred Marino
Design Collective

Lora Lee Martin
UC MBEST Center

Bud Mattingly
The Facility Group

Sherman McCorkle
Technology Ventures Corporation

Colin McCormick
U.S. House Science Committee

Sam McCormick
Sound Logic Advertising & Marketing, Inc.

Michael McCoy
Page Southerland Page, LLP

Mac McCullough
National Research Council

David McDonough
Johns Hopkins Real Estate

Jerry McGuire
University of North Carolina at Greensboro

Teresa McKnight
South Dakota State University
Innovation Campus

Robert McMahan
State of North Carolina

Marcia Mellitz
Center for Emerging Technologies

John Merrill
Gateway University Research Park

Keith Mock
Ballinger

Jackie Kerby Moore
Sandia Science and Technology Park

Francisco Moris
National Science Foundation

C. D. Mote, Jr.
University of Maryland

Murali Nair
National Science Foundation

Erik Necciai
Senate Small Business Committee

John Nguyen
Booz Allen Hamilton

John Niederhuber
National Cancer Institute

Alan O'Connor
RTI International

T. Olson
Olson & Associates Consulting, LLC

Victoria Palmer
Association of University Research
Parks

Diane Palmintera
Innovation Associates

Jaime Parada
Research and Innovation Technology Park

John Parks
University of South Carolina

Lisbeth Pettengill
Greater Baltimore Committee

Phillip H. Phan
Rensselaer Polytechnic Institute

Brian Phillips
Senate Small Business Committee

Steve Poteat
Montgomery College

Mazen Rasheed
Riyahd Techno Valley

Irene Redmiles
University of Maryland

Craig Reynolds
National Cancer Institute

Jon Riley
L&L Products, Inc.

Patricio Rojas
Fundacion Valle Lo Aguirre
Universidad de Chile

Wendy Schacht
Congressional Research Service

Lawrence Schuette
Office of Naval Research

Jane Shaab
UMB Research Park Corporation

Zhu Shen
BioForesight

Sujai Shivakumar
National Research Council

Harold Strong, Jr.
Texas A&M University Research Park

Tim Stuchtey
American Institute for Contemporary German Studies

Richard Stulen
Sandia National Laboratories

Jack Tillman
The University Financing Foundation

James Turner
U.S. House Science Committee

Tina Valdecanas
Research Triangle Park

Pablo Valenzuela
Fundacion Valle Lo Aguirre Universidad de Chile

Ilona Vass
Hungarian National Office for Research and Technology

Nicholas Vonortas
The George Washington University

Catherine Vorwald
Wexford Science & Technology, LLC

Eileen Walker
Association of University Research Parks

Rosemary Wander
University of North Carolina at Greensboro

Rick L. Weddle
Research Triangle Park

John Weete
Auburn Research Park

Charles Wessner
National Research Council

Kevin Wheeler
Senate Small Business Committee

Dave Wills
Ken Michaels Company

Alan Wm. Wolff
Dewey & LeBoeuf LLP

S. Pete Worden
NASA Ames Research Center

Bruce Wright
University of Arizona Office of Economic Development

Thomas Yang
Jensen & Partners

Ray Yin
Booz Allen Hamilton

Duncan York
Texas Research & Technology Foundation

Brian Young
Embassy of New Zealand

David Zumwalt
University of the Virgin Islands

Appendix C

Bibliography

Acs, Zoltan, and David Audretsch. 1990. *Innovation and Small Firms.* Cambridge, MA: The MIT Press.

Advanced Technology Program. 2001. *Performance of 50 Completed ATP Projects, Status Report Number 2.* NIST Special Publication 950-2. Washington, DC: U.S. Government Printing Office.

Advanced Technology Program and National Institute of Standards and Technology. 2003. *Beyond Measure: A Profile of ATP Health Care Investments.* Gaithersburg, MD: National Institute of Standards and Technology.

Advanced Technology Program and National Institute of Standards and Technology. 2003. *Powering Our High-Speed Economy: A Profile of ATP Energy Investments.* Gaithersburg, MD: National Institute of Standards and Technology.

Advanced Technology Program and National Institute of Standards and Technology. 2003. *Survey of ATP Applicants 2000.* NIST GCR 03-847. Gaithersburg, MD: National Institute of Standards and Technology.

Advanced Technology Program and National Institute of Standards and Technology. 2004. *Measuring ATP Impact: 2004 Report on Economic Progress.* Gaithersburg, MD: National Institute of Standards and Technology.

Advanced Technology Program and National Institute of Standards and Technology. 2005. *A Profile of ATP Manufacturing Investments, Inspiring Innovations in Industry.* Gaithersburg, MD: National Institute of Standards and Technology.

Advanced Technology Program and National Institute of Standards and Technology. 2005. *Survey of ATP Applicants 2002.* NIST GCR 05-876. Gaithersburg, MD: National Institute of Standards and Technology.

Aerts, Kris, and Dirk Czarnitzki. 2005. "Using Innovation Survey Data to Evaluate R&D Policy: The Case of Flanders." Catholic University of Leuven, Department of Applied Economics and Steunpunt O&O Statistieken.

Aerts, Kris, and Dirk Czarnitski. 2006. "The Impact of Public R&D Funding in Flanders." *IWT Studies* 54.

Aghion, Phillipe, Robin Burgess, Stephen Redding, and Fabrizio Zilibotti. 2003. "The Unequal Effects of Liberalization: Theory and Evidence from India." Washington, DC: Center for Economic Policy Research.

Agrawal, A. and R. Henderson. 2002. "Putting Patents in Context: Exploring Knowledge Transfer from MIT." *Management Science* 48(1):44-60.

Ahluwalia, Montek Singh. 2001. "State Level Reforms Under Economic Reforms in India." Stanford University Working Paper No. 96, March.

Aizcorbe, A., K. Flamm, and A. Kurshid. 2002. "The Role of Semiconductor Inputs in IT Hardware Price Decline: Computers vs. Communications." Federal Reserve Finance and Economics Discussion Paper 2002-37. Washington, DC: The Federal Reserve Board of Governors. August 2002; revised 2004.

Aizcorbe, A., S. Oliner, and D. Sichel. 2006. "Shifting Trends in Semiconductor Prices and the Pace of Technological Progress." Federal Reserve Board Finance and Economics Discussion Series Working Paper No. 2006-44. September.

Alberts, David, John Garska, and Frederick Stein. 1999. *Network Centric Warfare*. Washington, DC: Department of Defense Command and Control Research Program. Available at *<http://www.dodccrp.org/files/Alberts_NCW.pdf>*.

Alic, John A., Lewis M. Branscomb, Harvey Brooks, Ashton B. Carter, and Gerald L. Epstein. 1992. *Beyond Spin-off: Military and Commercial Technologies in a Changing World*. Boston, MA: Harvard Business School Press.

Allen, Stuart D., Albert N. Link, and Dan T. Rosenbaum. 2007. "Entrepreneurship and Human Capital: Evidence of Patenting Activity from the Academic Sector." *Entrepreneurship Theory and Practice* 31(6):937-951.

Allison, J., and M. Lemley. 1998. "Empirical Evidence on the Validity of Litigated Patents." *AIPLA Quarterly Journal* 26:185-277.

Altenburg, Tilman, Hubert Schmitz, and Andreas Stamm. 2008. "Breakthrough: China's and India's Transition from Production to Innovation." *World Development* 36(2):325-344.

American Intellectual Property Law Association. 2004. Reply to the National Academies Report. Washington, DC: American Intellectual Property Law Association.

Amsden, Alice H. 2001. *The Rise of "the Rest": Challenges to the West from Late-industrializing Economies*. Oxford, UK: Oxford University Press.

Amsden, Alice H. and Wan-wen Chu. 2003. *Beyond Late Development: Taiwan's Upgrading Policies*. Cambridge, MA: The MIT Press.

Amsden, Alice H., Ted Tschang, and Akira Goto. 2001. "Do Foreign Companies Conduct R&D in Developing Countries?" Tokyo, Japan: ADB Institute.

Anderson, Gary, Jeanne Powell, and Stephanie Shipp. 2003. "Improving the Advanced Technology Program's Business Reporting System—A Firm-Level R&D Survey." Proceedings of the American Statistical Association, Government Statistics Section, [CD-ROM]. Alexandria, VA: American Statistical Association.

Aoki, Reiko, and Sadao Nagaoka. 2004. "The Consortium Standard and Patent Pools." *The Economic Review (Keizai Kenkyu)* 55(4):345-356.

Aoki, Reiko, and Sadao Nagaoka. 2005. "Coalition Formation for a Consortium Standard through a Standard Body and a Patent Pool: Theory and Evidence from MPEG2, DVD and 3G." IIR Working Paper WP#05-01. February.

Applied Research Institute, Inc. 2006. *Survey of the Environment for Startups*. Applied Research Institute. November.

Archibugi, Danielle, Jeremy Howells, and Jonathan Michie, eds. 1999. *Innovation Policy and the Global Economy*. Cambridge, UK: Cambridge University Press.

Argyres, N. S., and J. P. Liebeskind. 1998. "Privatizing the Intellectual Commons: Universities and the Commercialization of Biotechnology." *Journal of Economic Behavior & Organization* 35:427-454.

Arora, A., and R. P. Merges. 2004. "Specialized Supply Firms, Property Rights, and Firm Boundaries." *Industrial and Corporate Change* 13(3):451-476.

Arora, A., M. Ceccagnoli, and W. Cohen. 2001. "R&D and the Patent Premium." Carnegie-Mellon University and INSEAD: paper presented at the ASSA Annual Meetings. January 2002. Atlanta, Georgia.

Arora, A., A. Fosfuri, and A. Gambardella. 2003. "Markets for Technology and Corporate Strategy." In O. Granstrand, ed., *Economics, Law, and Intellectual Property*, Boston, MA: Kluwer Academic Publishers.

Arrow K. J. 2000. "Increasing Returns: Historiographic Issues and Path Dependence." *European Journal of the History of Economic Thought* 7(2):171-180.

Arrow, Kenneth. 1962. *The Rate and Direction of Inventive Activity*. Princeton, NJ: Princeton University Press. Pp. 609-626.

Arthur, W. 1989. "Competing Technologies, Increasing Returns, and Lock-in by Historical Small Events." *Economic Journal* 99(2):116-131.

Asheim, Bjorn T. et al., eds. 2003. *Regional Innovation Policy for Small-medium Enterprises.* Cheltenham, UK: Edward Elgar.

Association of University Research Parks. 1998. "Worldwide Research & Science Park Directory 1998." New York: BPI Communications.

Association of University Research Parks. 2008 "The Power of Place: A National Strategy for Building America's Communities of Innovation." Tucson, AZ: Association of University Research Parks.

Athreye, Suma S. 2000. "Technology Policy and Innovation: The Role of Competition Between Firms." In Pedro Conceicao et al., eds. *Science, Technology, and Innovation Policy: Opportunities and Challenges for the Knowledge Economy*. Westport, CT: Quorum Books.

Atkinson, Robert. 2004. *The Past and Future of America's Economy–Long Waves of Innovation that Power Cycles of Growth*. Cheltenham, UK: Edward Elgar.

Atkinson, Robert. 2006. "Is the Next Economy Taking Shape?" *Issues in Science and Technology*. Winter.

Audretsch, D. B., ed. 1998. *Industrial Policy and Competitive Advantage*, Volumes 1 and 2. Cheltenham, UK: Edward Elgar.

Audretsch, D. B. 1998. "Agglomeration and the Location of Innovative Activity." *Oxford Review of Economic Policy* 14(2):18-29.

Audretsch, D. B. 2001. "The Prospects for a Technology Park at Ames: A New Economy Model for Industry-Government Partnership?" In National Research Council. *A Review of the New Initiatives at the NASA Ames Research Center*. Charles W. Wessner, ed. Washington, DC: National Academy Press.

Audretsch, D. B. 2006. *The Entrepreneurial Society*. Oxford, UK: Oxford University Press.

Audretsch, D. B., and M. P. Feldman. 1996. "R&D Spillovers and the Geography of Innovation and Production." *American Economic Review* 86(3):630-640.

Audretsch, D. B., and M. P. Feldman. 1999. "Innovation in Cities: Science-based Diversity, Specialization, and Localized Competition." *European Economic Review* 43(2):409-429.

Audretsch, D. B., B. Bozeman, K. L. Combs, M. P. Feldman, A. N. Link, D. S. Siegel, P. Stephan, G. Tassey, and C. Wessner. 2002. "The Economics of Science and Technology." *Journal of Technology Transfer* 27:155-203.

Audretsch, D. B., H. Grimm, and C. W. Wessner. 2005. *Local Heroes in the Global Village: Globalization and the New Entrepreneurship Policies.* New York: Springer.

Auerswald, Philip E., Lewis M. Branscomb, Nicholas Demos, and Brian K. Min. 2005. *Understanding Private-Sector Decision Making for Early-Stage Technology Development: A "Between Invention and Innovation Project" Report.* NIST GCR 02–841A. Gaithersburg, MD: National Institute of Standards and Technology.

Augustine, Norman. 2007. *Is America Falling Off the Flat Earth?* Washington, DC: The National Academies Press.

Baker, Stephen. 2005. "New York's Big Hopes for Nano." *BusinessWeek* February 4.

Bakouros, Y. L., D. C. Mardas, and N. C. Varsakelis. 2002. "Science Parks, a High-Tech Fantasy? An Analysis of the Science Parks of Greece." *Technovation* 22(2):123-128.

Baldwin, J. R., P. Hanl, and D. Sabourin. 2000. "Determinants of Innovative Activity in Canadian Manufacturing Firms: The Role of Intellectual Property Rights." Statistics Canada Working Paper No. 122. March.

Baldwin, John Russel, and Peter Hanel. 2003. *Innovation and Knowledge Creation in an Open Economy: Canadian Industry and International Implications*. Cambridge, UK: Cambridge University Press.

Balzat, Markus, and Andreas Pyka. 2006. "Mapping National Innovation Systems in the OECD Area." *International Journal of Technology and Globalisation* 2(1-2):158-176.

Baptista, R. 1998. "Clusters, Innovation, and Growth: A Survey of the Literature." In G. M. P. Swann, M. Prevezer, and D. Stout, eds. *The Dynamics of Industrial Clustering*. Oxford, UK: Oxford University Press.

Bartzokas, Anthony, and Morris Teubal. 2002. "The Political Economy of Innovation Policy Implementation in Developing Countries." *Economics of Innovation and New Technology* 11(4-5).

Beinhocker, Eric D. 2007. *Origin of Wealth—Evolution, Complexity, and the Radical Remaking of Economics*. Cambridge, MA: Harvard Business School.

Bennis, Warren, and Patricia Ward Biederman. 1997. *Organizing Genius*. New York: Basic Books.

Berlin, Leslie. 2005. *The Man Behind the Microchip: Robert Noyce and the Invention of Silicon Valley.* New York: Oxford University Press.

Bessen, J., and M. J. Meurer. 2005. "The Patent Litigation Explosion," Research on Innovation and Boston University School of Law: manuscript. August.

Bhidé, Amar. 2006. "Venturesome Consumption, Innovation and Globalization." Paper presented at the Centre on Capitalism & Society and CESifo Venice Summer Institute 2006, "Perspectives on the Performance of the Continent's Economies," 21-22 July 2006. Held at Venice International University. San Servolo, Italy.

Biegelbauer, Peter S., and Susana Borras, eds. 2003. *Innovation Policies in Europe and the U.S.: The New Agenda*. Aldershot, UK: Ashgate.

Birch, David. 1981. "Who Creates Jobs?" *The Public Interest* 65:3-14.

Blanpied, William A. 1998. "Inventing U.S. Science Policy." *Physics Today* 51(2):34-40.

Block, Fred, and Matthew Keller. 2008. "Where Do Innovations Come From? Transformations in the U.S. National Innovation System, 1970-2006." The Information Technology and Innovation Forum. July.

Blomström, Magnus, Ari Kokko, and Fredrik Sjöholm. 2002. "Growth & Innovation Policies for a Knowledge Economy: Experiences from Finland, Sweden, & Singapore." EIJS Working Paper, Series No. 156.

Bloomberg News. 2006. "The Next Green Revolution." August 21.

Bonvillian, William B. 2006. "Power Play, The DARPA Model and U.S. Energy Policy." *The American Interest* II(2):39-48.

Booz Allen Hamilton. 2005. "Booz Allen Hamilton Global Innovation 1000: Money Isn't Everything." *Strategy + Business,* 41.

Borras, Susana. 2003. *The Innovation Policy of the European Union: From Government to Governance.* Cheltenham, UK: Edward Elgar.

Borrus, Michael, and Jay Stowsky. 2000. "Technology Policy and Economic Growth." In Charles Edquist and Maureen McKelvey, eds. *Systems of Innovation: Growth, Competitiveness and Employment, Vol. 2*. Cheltenham, UK: Edward Elgar.

Branscomb, L. M., and P. E. Auerswald. 2001. *Taking Technical Risks: How Innovators, Executives, and Investors Manage High-Tech Risks*. Boston, MA: The MIT Press.

Branscomb, Lewis M., and Philip E. Auerswald. 2002. *Between Invention and Innovation: An Analysis of Funding for Early-Stage Technology Development*. NIST GCR 02–841. Gathersburg, MD: National Institute of Standards and Technology. November.

Braudel, Fernand. 1973. *Capitalism and Material Life 1400-1800*. London, UK: Harper Colophon Books.

Breschi, S. and F. Lissoin. 2001. "Knowledge Spillovers and Local Innovation Systems: A Critical Survey." *Industrial and Corporate Change* 10(4):975-1005.

Browning, L., and J. Shetler. 2000. *SEMATECH: Saving the U.S. Semiconductor Industry*. College Station, TX: Texas A&M University Press.

Buchanan, James M. 1987. "An Economic Theory of Clubs." In *Economics: Between a Predictive Science and Moral Philosophy*. College Station, TX: Texas A&M University Press.

Bush, Nathan. 2005. "Chinese Competition Policy, It Takes More than a Law." *China Business Review*. May-June.

Bush, Vannevar. 1945. *Science: The Endless Frontier.* Washington, DC: U.S. Government Printing Office.

Capron, Henri, and Michele Cincera. 2006. *Strengths and Weaknesses of the Flemish Innovation System: An External Viewpoint.* Brussels, Belgium: IWT.

Caracostas, Paraskevas, and Ugur Muldur. 2001. "The Emergence of the New European Union Research and Innovation Policy." In P. Laredo and P. Mustar, eds. *Research and Innovation Policies in the New Global Economy: An International Comparative Analysis*. Cheltenham, UK: Edward Elgar.

Castells, M., and P. Hall. 1994. *Technopoles of the World*. London, UK: Oxford University Press.

Cebrowski, Arthur, and John Garska. 1998. "Network Centric Warfare: Its Origin and Future." *U.S. Naval Institute Proceedings*. January.

Chambers, John, ed. 1999. *The Oxford Companion to American Military History.* Oxford, UK: Oxford University Press.

Chan, K. F., and Theresa Lau. 2005. "Assessing Technology Incubator Programs in the Science Park: The Good, the Bad and the Ugly." *Technovation* 25(10):1215-1228.

Chand, Satish, and Kunal Sen. 2002. "Trade Liberalization and Productivity Growth: Evidence from Indian Manufacturing." *Review of Development Economics* 6, February.

Chang, Connie, Stephanie Shipp, and Andrew Wang. 2002. The Advanced Technology Program: A Public-Private Partnership for Early-stage Technology Development. *Venture Capital* 4(4): 363-370.

Chesbrough, Henry. 2003. *Open Innovation: The New Imperative for Creating and Profiting from Technology*. Cambridge, MA: Harvard Business School Press. April.

China Daily. 2008. "China Luring 'Sea Turtles' Home." December 18.

China Trade Extra. 2005. *China Agrees to Delay Software Procurement Rule While Talking with U.S.* July 11.

Chinese Ministry of Finance. 2006. *Opinions of the Ministry of Finance on Implementing Government Procurement Policies That Encourage Indigenous Innovation*. Cai Ku [2006] No. 47, June 13.

Chordà, I. M. 1996. "Towards the Maturity State: An Insight into the Performance of French Technopoles." *Technovation* 16(3):143-152.

Chuma, Hiroyuki. 2006. "Increasing Complexity and Limits of Organization in the Microlithography Industry: Implications for Science-based Industries." *Research Policy* 35:394-411.

Chuma, Hiroyuki, and Norikazu Hashimoto. 2007. "Moore's Law, Increasing Complexity and Limits of Organization: Modern Significance of Japanese DRAM ERA." NISTEP Discussion Paper No. 44. National Institute of Science and Technology Policy.

Cimoli, Mario, and Marina della Giusta. 2000. "The Nature of Technological Change and its Main Implications on National and Local Systems of Innovation." IIASA Interim Report IR-98-029.

Cincera, Michele. 2006. *Comparison of Regional Approaches to Foster Innovation in the European Union: The Case of Flanders*. Brussels, Belgium: IWT.

Cincera, Michele. 2006. *R&D Activities of Flemish Companies in the Private Sector: An Analysis for the Period 1998-2002*. Brussels, Belgium: IWT.

Clark, B. 1995. *Places of Inquiry*. Berkeley, CA: University of California Press.

Clarke, P. 2004. "LETI, Crolles Alliance Open $350-million 32-nm Research Fab." *EE Times* April 24.

Clarke, P., M. LaPedus, and M. Santarini. 2005. "IBM-led Consortium to Build Fab in N.Y." *EE Times* January 5.

Clough, G. Wayne. 2007. "The Role of the Research University in Fostering Innovation." The Americas Competitiveness Forum. June 12.

Coakes, Elayne, and Peter Smith. 2007. "Developing Communities of Innovation by Identifying Innovation Champions." *The Learning Organization: An International Journal* 14(1):74-85.

Coburn, Christopher, and Dan Berglund. 1995. *Partnerships: A Compendium of State and Federal Cooperative Programs*. Columbus, OH: Battelle Press.

Cohen, W. 2002. "Thoughts and Questions on Science Parks." Presented at the National Science Foundation Science Parks Indicators Workshop. University of North Carolina at Greensboro.

Cohen, W., R. Florida, and R. Goe. 1992. *University-Industry Research Centers in the United States*. Pittsburgh, PA: Carnegie-Mellon University.

Cohen, W., R. Nelson, and J. Walsh. 2000. "Protecting Their Intellectual Assets: Appropriability Conditions and Why U.S. Manufacturing Firms Patent (or Not)." NBER Working Paper 7552. Cambridge, MA: National Bureau of Economic Research.

Cohen, W. M., A. Goto, A. Nagata, R. R. Nelson, and J. P. Walsh. 2002. "R&D Spillovers, Patents and the Incentives to Innovate in Japan and the United States." *Research Policy* 1425:1-19.

Combs, Kathryn L., and Albert N. Link. 2003. "Innovation Policy in Search of an Economic Paradigm: The Case of Research Partnerships in the United States." *Technology Analysis & Strategic Management* 15(2).

Conant, Jennet. 2002. *Tuxedo Park*. New York: Simon & Shuster.

Conant, Jennet. 2005. *109 East Palace*. New York: Simon & Shuster.

Council on Competitiveness. 2005. *Innovate America: Thriving in a World of Challenge and Change*. Washington, DC: Council on Competitiveness.

Council on Government Relations. 2000. *Technology Transfer in U.S. Research Universities: Dispelling Common Myths*. Washington, DC: Council on Government Relations.

Council for Science and Technology Policy. 2002. *Reports on the Reform for Competitive Research Grants*. (in Japanese).

Council for Science and Technology Policy. 2003. *Reform of Competitive Funding System*. (*Kyousouteki Sikinseido no kaikaku*). Tokyo, Japan: Council for Science and Technology Policy. April 15.

Council for Science and Technology Policy. 2005. *Reports of the Expert Panel on the Basic Policy*. *<http://www8.cao.go.jp/cstp/english/index.html>*.

Council for Science and Technology Policy. 2005. *Report on S&T Basic Policy*. (*Kagaku-gijutu-kihon-seisaku-ni-tsuite ni-taisuru tousin*). Tokyo, Japan: Council for Science and Technology Policy. December 27.

Crafts, N. F. R. 1995. "The Golden Age of Economic Growth in Western Europe, 1950-1973." *Economic History Review* 3:429-447.

Czarnitzki, Dirk, and Niall O'Byrnes. 2007. "Innovation and the Impact on Productivity in Flanders." *Tijdschrift voor Economie en Management* 52(2).

Dahlman, Carl J. 2005. *India and the Knowledge Economy: Leveraging Strengths and Opportunities*. Washington, DC: World Bank Publications.

Dahlman, Carl J., and Jean Eric Aubert. 2001. *China and the Knowledge Economy: Seizing the 21st Century*. Washington, DC: The World Bank.

Dahlman, Carl, and Anuja Utz. 2005. *India and the Knowledge Economy: Leveraging Strengths and Opportunities*. Washington, DC: The World Bank.

Daneke, Gregory A. 1998. "Beyond Schumpeter: Non-linear Economics and the Evolution of the U.S. Innovation System." *Journal of Socio-economics* 27(1):97-117.

Darby, Michael, Lynne G. Zucker, and Andrew J. Wang. 2002. *Program Design and Firm Success in the Advanced Technology Program: Project Structure and Innovation Outcomes*. NISTIR 6943. Gaithersburg, MD: National Institute of Standards and Technology.

Das, Gurcharan. 2006. "The India Model." *Foreign Affairs* 85(4).

Dasgupta, P., and P. David. 1994. "Toward a New Economics of Science." *Research Policy* 23: 487-521.

Davidsson, Per. 1996. "Methodological Concerns in the Estimation of Job Creation in Different Firm Size Classes." Working Paper, Jönköping International Business School.

David, P. A. 1985. "Clio and the Economics of QWERTY." *American Economic Review* 75(2):332-337.

Davis, Steven, John Haltiwanger, and Scott Schuh. 1993. "Small Business and Job Creation: Dissecting the Myth and Reassessing the Facts." Working Paper No. 4492. Cambridge, MA: National Bureau of Economic Research.

Debackere, Koenraad, and Wolfgang Glänzel. 2004. "Using a Bibliometric Approach to Support Research Policy Making: The Case of the Flemish BOF-key." *Scientometrics* 59(2).

Debackere, Koenraad, and Reinhilde Veugelers. 2005. "The Role of Academic Technology Transfer Organizations in Improving Industry Science Links." *Research Policy* 34(3):321-342.

Defense Advanced Research Projects Agency. 2003. *DARPA Over The Years*. Arlington, VA: Defense Advanced Research Projects Agency. October 27.

Defense Advanced Research Projects Agency. 2005. *DARPA—Bridging the Gap, Powered by Ideas*. Arlington, VA: Defense Advanced Research Projects Agency. February.

Defense Science Board. 2005. *High Performance Microchip Supply*. February.

de Jonquieres, Guy. 2004. "China and India Cannot Fill the World's Skills Gap." *Financial Times* July 12.

de Jonquieres, Guy. 2004. "To Innovate, China Needs More than Standards." *Financial Times* July 12.

de Jonquieres, Guy. 2006. "China's Curious Marriage of Convenience." *Financial Times* July 19.

De la Mothe, J., and Gilles Paquet. 1998. "National Innovation Systems, 'Real Economies' and Instituted Processes." *Small Business Economics* 11:101-111.

De Proft, A. 2006. Presentation at symposium on "Synergies in Regional and National Policies in the Global Economy." Leuven, Belgium. September.

Doloreux, David. 2004. "Regional Innovation Systems in Canada: A Comparative Study." *Regional Studies* 38(5):479-492.

Domain-b.com. 2005. "Nasscom-McKinsey: India to Face Skilled Workers' Shortage by Next Decade." December 12. Available online at *<http://www.domain-b.com/organisation/nasscom/20051217_shortage.html>*.

Dries, Ilse, Peer van Humbeek, and Jan Larosse. 2005. "Linking Innovation Policy and Sustainable Development in Flanders." Paris, France: Organisation for Economic Co-operation and Development.

Dudas, J. 2005. "Statement of the Honorable Jon W. Dudas Deputy Under Secretary of Commerce for Intellectual Property and Director of the U.S. Patent and Trademark Office before the Subcommittee on Intellectual Property, Committee on the Judiciary." U.S. Senate. *<http://judiciary.senate.gov>*.

Eaton, Jonathan, Eva Gutierrez, and Samuel Kortum. 1998. "European Technology Policy." NBER Working Papers 6827.

The Economist. 2005. "Competing Through Innovation." December 17.

Edler, J., and S. Kuhlmann. 2005. "Towards One System? The European Research Area Initiative, the Integration of Research Systems and the Changing Leeway of National Policies." *Technikfolgenabschätzung: Theorie und Praxis* 1(4):59-68.

EE Times. 2006. "Chinese Province Pays to Get 300-mm Wafer Fab." June 28.

Ehlen, Mark A. 1999. *Economic Impacts of Flow-Control Machining Technologies: Early Applications in the Automobile Industry*. NISTIR 6373. Gaithersburg, MD: National Institute of Standards and Technology.

Ehlers, Vernon J. 1998. *Unlocking Our Future: Toward a New National Science Policy, A Report to Congress by the House Committee on Science.* Washington, DC: U.S. Government Printing Office.

Ehrenberg, R. 2000. *Tuition Rising*. Cambridge, MA: Harvard University Press.

Eickelpasch, Alexander, and Michael Fritsch. 2005. "Contests for Cooperation: A New Approach in German Innovation Policy." *Research Policy* 34:1269-1282.

Electronic News. 2006. "SMIC Gets $3B Nod from Chain's Wuhan Government." May 22.

Endquist, Charles, ed. 1997. *Systems of Innovation: Technologies, Institutions, and Organizations*. London, UK: Pinter.

EOS Gallup Europe. 2004. *Entrepreneurship.* Flash Eurobarometer 146. January. Accessed at <*http://ec.europa.eu/enterprise/enterprise_policy/survey/eurobarometer146_en.pdf*>.

European Commission. 2003. *Innovation in Candidate Countries: Strengthening Industrial Performance.* Luxembourg: Office for Official Publications of the European Communities, May.

European Commission. 2003. "Investing in Research: An Action Plan for Europe 2003." Luxembourg: Office for Official Publications of the European Communities.

European Commission. 2003. *3rd S&T Indicators Report*. Luxembourg: Office for Official Publications of the European Communities.

Evans, D., and Jovanovic, B. 1989. "An Estimated Model of Entrepreneurial Choice under Liquidity Constraints." *Journal of Political Economy* 97:808-827.

Evans, Sir Harold. 2005. *They Made America*. Sloan Foundation Project. New York: Little, Brown and Company.

Faems, Dries, Bart Van Looy, and Koenraad Debackere. 2005. "Inter-organizational Collaboration and Innovation: Toward a Portfolio Approach." *Journal of Product Innovation Management* 22(3):238-250.

Fan, W., and White, M. J. 2002. "Personal Bankruptcy and the Level of Entrepreneurial Activity." NBER Working Paper, Series 9340. Boston, MA: National Bureau of Economic Research.

Fangerberg, Jan. 2002. *Technology, Growth, and Competitiveness: Selected Essays*. Cheltenham, UK: Edward Elgar.

Federal Register Notice. 2004. "2004 WTO Dispute Settlement Proceeding Regarding China: Value-Added Tax on Integrated Circuits." April 21.

Federal Trade Commission. 2003. "To Promote Innovation: The Proper Balance of Patent and Competition Law Policy." Washington, DC: U.S. Government Printing Office. October.

Feldman, Maryann, and Maryellen Kelley. 2001. *Winning an Award from the Advanced Technology Program*. NISTIR-6577. Gaithersburg, MD: National Institute of Standards and Technology.

Feldman, Maryann and Albert N. Link. 2001. "Innovation Policy in the Knowledge-based Economy." *Economics of Science, Technology and Innovation Volume 23*. Boston, MA: Kluwer Academic Press.

Feldman, Maryann P., Albert N. Link, and Donald S. Siegel. 2002. *The Economics of Science and Technology: An Overview of Initiatives to Foster Innovation, Entrepreneurship, and Economic Growth.* Boston, MA: Kluwer Academic Press.

Feldman, Maryann, Irwin Feller, Janet Bercovitz, and Richard Burton. 2002. "Equity and the Technology Transfer Strategies of American Research Universities." *Management Science* 48(1):105-121.

Feller, Irwin. 1994. "University-Industry R&D Relationships." In J. Schmandt and R. Wilson, eds., *Growth Policy in the Age of High Technology*. Boston, MA: Unwin Hyman.

Feller, Irwin. 1997. "Technology Transfer from Universities." In John Smart, ed., *Higher Education: Handbook of Theory and Research, Vol. XII*. New York: Agathon Press.

Feller, Irwin. 2004. *A Comparative Analysis of the Processes and Organizational Strategies Engaged in by Research Universities Participating in Industry-University Research Relationships*. Final report submitted to the University of California Industry-University Cooperative Research Program. Agreement No. M-447646-19927-3.

Ferguson, R. and C. Olofsson. 2004. "Science Parks and the Development of NTBFs: Location, Survival and Growth." *Journal of Technology Transfer* 29(1): 5-17.

Field, A. J. 2003. "The Most Technologically Progressive Decade of the Century." *American Economic Review* (September):1406.

The Financial Times. 2005. "World Leader in Patents Focuses on Incremental Innovations." October 12.

The Financial Times. 2006. "India Needs Big Infrastructure Drive." February 23.

Flamm, K. 1996. *Mismanaged Trade: Strategic Policy and the Semiconductor Industry*. Washington, DC: The Brookings Institution.

Flamm, K. 2003. "Microelectronics Innovation: Understanding Moore's Law and Semiconductor Price Trends." *International Journal of Technology, Policy, and Management* 3(2).

Flamm, K. 2003. "The New Economy in Historical Perspective: Evolution of Digital Technology." In *New Economy Handbook*. St. Louis, MO: Academic Press.

Flamm, Kenneth. 2003. "SEMATECH Revisited: Assessing Consortium Impacts on Semiconductor Industry R&D." In National Research Council. *Securing the Future: Regional and National Programs to Support the Semiconductor Industry*. Charles W. Wessner, ed. Washington, DC: The National Academies Press.

Flamm, K. 2006. "The Economics of Innovation in Microprocessors." Draft. December.

Florida, Richard, Tim Gulden, and Charlotta Mellander. 2007. "The Rise of the Mega-Region." October.

Foerst, Ann. 2005. *God in the Machine*. New York: Penguin Books.

Fogarty, Michael S., Amit K. Sinha, and Adam B. Jaffe. Forthcoming. *ATP and the U.S. Innovation System—A Methodology for Identifying Enabling R&D Spillover Networks with Applications to Micro-electromechanical Systems (MEMS) and Optical Recording*. Draft report submitted to ATP.

Fonfria, Antonio, Carlos Diaz de la Guardia, and Isabel Alvarez. 2002. "The Role of Technology and Competitiveness Policies: A Technology Gap Approach." *Journal of Interdisciplinary Economics* 13(1-2-3):223-241.

Fong, Glenn R. 2001. "ARPA Does Windows: The Defense Underpinning of the PC Revolution," *Business and Politics* 3(3).

Foray, Dominique, and Patrick Llerena. 1996. "Information Structure and Coordination in Technology Policy: A Theoretical Model and Two Case Studies." *Journal of Evolutionary Economics* 6(2).

Fortune. 2006. "America's Most Admired Companies 2006."

Friedman, Thomas. 2005. *The World Is Flat: A Brief History of the 21st Century*. New York: W. H. Freeman.

Fujimura, Shuzo. 2000. "Handotai Rikkoku Futatabi." (In Japanese.) Nikkann-Kogyo Shin-bun-sha.

Fujimura, Shuzo, and Hiroyuki Chuma. 2006. "Semiconductor Consortia in Japan: Experiences and Lessons." Presentation at the conference, *21st Century Innovation Systems for the United States and Japan: Lessons from a Decade of Change*. Tokyo, Japan: The National Academies, NISTEP of Japan, and The Institute of Innovation Research of Hitotsubashi University. *<http://www.nistep.go.jp/IC/ic060110/pdf/5-2.pdf>*.Fukao, Kuoji, and Hyeog Ug Kwon. 2005. "Why Did Japan's TFP Growth Slow Down in the Lost Decade? An Empirical Analysis Bases on Firm-Level Data of Manufacturing Firms" *RIETI Discussion Paper Series*. 05-E-004.

Fukugawa, N. 2006. "Science Parks in Japan and Their Value-Added Contributions to New Technology-based Firms" *International Journal of Industrial Organization* 24(2):381-400.

Furman, Jeffrey L., Michael E. Porter, and Scott Stern. 2002. "The Determinants of National Innovative Capacity." *Research Policy* 31:899-933.

Gallaher, M. P., A. N. Link, and J. E. Petrusa. 2006. *Innovation in the U.S. Service Sector*. London, UK: Routledge.

Geiger, R. 1986. *To Advance Knowledge*. New York: Oxford University Press.

Geiger, R. 1993. *Research and Relevant Knowledge.* New York: Oxford University Press.

George, Gerard, and Ganesh N. Prabhu. 2003. "Developmental Financial Institutions as Technology Policy Instruments: Implications for Innovation and Entrepreneurship in Emerging Economies." *Research Policy* 32(1):89-108.

Gibb, M. J. 1985. *Science Parks and Innovation Centres: Their Economic and Social Impact*. Amsterdam, The Netherlands: Elsevier.

Goldstein, H. A., and M. I. Luger. 1990. "Science/Technology Parks and Regional Development Theory." *Economic Development Quarterly* 4(1):64-78.

Goldstein, H. A., and M. I. Luger. 1992. "University-based Research Parks as a Rural Development Strategy." *Policy Studies Journal* 20(2):249-263.

Good, Mary. 2005. Presentation at the National Academies Conference, "Accelerating Innovation." Washington, DC. October 19.

Goodwin, James C., et al. 1999. *Technology Transition*. Arlington, VA: Defense Advanced Research Projects Agency.

Government of the People's Republic of China. 2004. *Anticompetitive Practices of Multinational Companies and Countermeasures*. Administration of Industry and Commerce, Office of Anti-monopoly, Fair Trade Bureau, State Administration of Industry and Commerce. May.

Government of the People's Republic of China, Office of Anti-monopoly, Fair Trade Bureau, State Administration of Industry and Commerce. 2004. "Anticompetitive Practices of Multinational Companies in China and Countermeasures." *Journal of Administration of Industry and Commerce.* May.

Government of the People's Republic of China. 2005. *Article 10: Exemptions of Monopoly Agreements*. Anti-Monopoly Law of the People's Republic of China. Revised July 27.

Government of the People's Republic of China, Ministry of Information Industry. 2006. "Outline of the 11th Five-Year Plan and Medium-and-Long-Term Plan for 2020 for Science and Technology Development in the Information Industry." Xin Bu Ke, No. 309, August 29.

Government of the People's Republic of China, National Development and Reform Commission. 2006. The 11th Five-Year Plan. March 19. *<http://english.gov.cn/2006-07/26/content_346731.htm>*.

Grande, Edgar. 2001. "The Erosion of State Capacity and European Innovation Policy: A Comparison of German and EU Information Technology Policies." *Research Policy* 30(6):905-921.

Grayson, L. 1993. *Science Parks: An Experiment in High-Technology Transfer* London, UK: The British Library Board.

Griffing, Bruce. 2001. "Between Invention and Innovation, Mapping the Funding for Early-Stage Technologies." Presentation at Carnegie Conference Center. Washington, DC. January 25.

Griliches, Z. 1993. "The Search for R&D Spillovers." *Scandinavian Journal of Economics* 94(S): S29-S47.

Grindley, Peter, David Mowery, and Brian Silverman. 1994. "SEMATECH and Collaborative Research: Lessons in the Design of High Technology Consortia." *Journal of Policy Analysis and Management* 13(4):723-758.

Grove, A. 2001. *Swimming Across.* New York: Warner Books.

Guellec, Dominique, and Bruno van Pottelsberghe de la Potterie. 2001. "R&D and Productivity Growth: Panel Data Analysis of 16 OECD Countries." Organisation for Economic Co-operation and Development. DSTI Working Papers 2001/314. June.

Guo Ban Han No. 30. 2006. Letter from the General Office of the State Council on Approving the Formulation of the Rules for Implementation of the Several Supporting Policies for Implementation of the Outline of the National Medium and Long-term Plan for Development of Science and Technology. *Gazette of the State Council*. Issue No. 17, Serial No. 1196, June 20.

Guy, I. 1996. "A Look at Aston Science Park." *Technovation* 16(5):217-218.

Guy, I. 1996. "New Ventures on an Ancient Campus." *Technovation* 16(6):269-270.

Hackett, S. M., and D. M. Dilts. 2004. "A Systematic Review of Business Incubation Research." *Journal of Technology Transfer* 29(1):55-82.

Hall, B. H. 2002. "The Assessment: Technology Policy." *Oxford Review of Economic Policy* 18(1):1-9.

Hall, B. H. 2004. "Exploring the Patent Explosion." NBER Working Paper 10605. Cambridge, MA: National Bureau of Economic Research.

Hall, B. H. 2005. "Exploring the Patent Explosion." *Journal of Technology Transfer* 30:35-48.

Hall, B. H., and R. H. Ziedonis. 2001. "The Patent Paradox Revisited: An Empirical Study of Patenting in the U.S. Semiconductor Industry. 1979-1995." *RAND Journal of Economics* 32:101-128.

Hall, B. H., A. N. Link, and J. T. Scott. 2001. "Barriers Inhibiting Industry from Partnering with Universities: Evidence from the Advanced Technology Program." *Journal of Technology Transfer* 26(1-2):87-98.

Hall, B. H., A. N. Link, and J. T. Scott. 2003. "Universities as Research Partners." *Review of Economics and Statistics* 85(2):485-491.

Hall, B. H., A. Jaffe, and M. Trajtenberg. 2005. "Market Value and Patent Citations." *RAND Journal of Economics* 36:16-38.

Halpern, Mark. 2006. "The Trouble with the Turing Test." *The New Atlantis* 11(Winter):42-63.

Hane, G. 1999. "Comparing University-Industry Linkages in the United States and Japan." In L. Branscomb, F. Kodama, and R. Florida, eds. *Industrializing Knowledge*. Cambridge, MA: The MIT Press.

Hansson, F., K. Husted, and J. Vestergaard. 2005. "Second Generation Science Parks: From Structural Holes Jockeys to Social Capital Catalysts of the Knowledge Society." *Technovation* 25(9):1039-1049.

Hart, David. 1998. *Forged Consensus*. Princeton, NJ: Princeton University Press.

Hashimoto, T. 1999. "The Hesitant Relationship Reconsidered: University-Industry Cooperation in Postwar Japan." In Lewis M. Branscomb, Fumio Kodama, and Richard Florida, eds. *Industrializing Knowledge: University-Industry Linkage in Japan and the United States*. Cambridge, MA: The MIT Press.

Hayashi, Fumio, and Edward C. Prescott. 2002. "The 1990s in Japan: A Lost Decade." *Review of Economic Dynamics* 5(1):206-235.

Heller, M., and R. Eisenberg. 1998. "Can Patents Deter Innovation? The Anticommons in Biomedical Research." *Science* 280:698.

Henderson, J. V. 1986. "The Efficiency of Resource Usage and City Size." *Journal of Urban Economics* 19(1):47-70.

Henderson, Jennifer A., and John J. Smith. 2002. "Academia, Industry, and the Bayh-Dole Act: An Implied Duty to Commercialize." White Paper. Center for the Integration of Medicine and Innovative Technology. Harvard University. October.

Henderson, R., M. Trajtenberg, and A. Jaffe. 1998. "Universities as a Source of Commercial Technology: A Detailed Analysis of University Patenting, 1965-1988." *Review of Economics and Statistics* 80(1):119-127.

Henderson, R., L. Orsenigo, and G. P. Pisano. 1999. "The Pharmaceutical Industry and the Revolution in Molecular Biology: Interactions among Scientific, Institutional, and Organizational Change." In D. C. Mowery and R. R. Nelson, eds. *Sources of Industrial Leadership*. Cambridge, UK: Cambridge University Press.

Hennessy, J. L., and D. A. Patterson 2002. *Computer Architecture: A Quantitative Approach*. 3rd edition. San Francisco, CA: Morgan Kaufmann Publishers Inc.

Hicks, D., T. Breitzman, D. Olivastro, and K. Hamilton. 2001. "The Changing Composition of Innovative Activity in the U.S.—A Portrait Based on Patent Analysis." *Research Policy* 30(4):681-704.

Hilpert, U., and B. Ruffieux. 1991. "Innovation, Politics and Regional Development: Technology Parks and Regional Participation in High-Technology in France and West Germany." In U. Hilpert, ed. *Regional Innovation and Decentralization: High-Technology Industry and Government* Policy. London, UK: Routledge.

Himmelberg, C., and B. C. Petersen. 1994. "R&D and Internal Finance: A Panel Study of Small Firms in High-Tech Industries." *Review of Economics and Statistics* 76(1):38-51.

Holtz-Eakin, D., D. Joulfian, and H. Rosen. 1994. "Striking it Out: Entrepreneurial Survival and Liquidity Constraints." *Journal of Political Economy* 102(1):53-75.

Hoshi, Takeo, and Anil Kashyap. 2004. "Japan's Economic and Financial Crisis: An Overview." *The Journal of Economic Perspectives* Winter.

Howell, Thomas. 2003. "Competing Programs: Government Support for Microelectronics." In National Research Council. *Securing the Future: Regional and National Programs to Support the Semiconductor Industry.* Charles W. Wessner, ed. Washington, DC: The National Academies Press.

Howell, Thomas, et al. 2003. *China's Emerging Semiconductor Industry*. San Jose, CA: Semiconductor Industry Association, October.

Hu, Zhijian. 2006. "IPR Policies In China: Challenges and Directions." Presentation at *Industrial Innovation in China*. Levin Institute Conference. July 24-26.

Huang, Yasheng, and Tarun Khanna. 2003. "Can India Overtake China?" *Foreign Policy* July-August.

Huddleson, Lillian, and Vicki Daitch. 2002. *True Genius—The Life and Science of John Bardeen*. Washington, DC: Joseph Henry Press.

Hughes, Kent. 2005. *Building the Next American Century: The Past and Future of American Economic Competitiveness*. Washington, DC: Woodrow Wilson Center Press. Chapter 14.

Hughes, Kent H. 2005. "Facing the Global Competitiveness Challenge." *Issues in Science and Technology* XXI(4):72-78.

Hulsink, W., H. Bouwman, and T. Elfring. 2007. "Silicon Valley in the Polder? Entrepreneurial Dynamics, Virtuous Clusters and Vicious Firms in the Netherlands and Flanders." ERIM Report Series Research in Management (ERS-2007-048-ORG).

Hundley, Richard O. 1999. *Past Revolutions, Future Transformations: What Can the History of Revolutions in Military Affairs Tell Us About Transforming the U.S. Military*. RAND Corporation, National Research Institute.

Iansati, Marco, and Roy Levien. 2005. *The Keystone Advantage*. Cambridge, MA: Harvard Business School Press.

Industrial Research Institute, Inc. 2001. "Industry-University Intellectual Property." Position Paper. External Research Directors Network. April.

Inkpen, Andrew C. and Wang Pien. 2006. "An Examination of Collaboration and Knowledge Transfer: China–Singapore Suzhou Industrial Park." *Journal of Management Studies* 43(4):779-811.

Inside U.S.-China Trade. 2006. *Industry Worried China Backing out of Commitment to Join GPA*. September 27.

International Association of Science Parks (IASP). 2000. *<http://www.iaspworld.org/information/definitions.php>*.

International Herald Tribune. 2006. "Start-Ups Explore Abroad for IPOs." December 25.

Jacobs, Tom. 2002. "Biotech Follows Dot.com Boom and Bust." *Nature* 20(10):973.

Jaffe, A. B. 1989. "Real Effects of Academic Research." *American Economic Review* 79(5): 957-970.

Jaffe, A. B. 1997. *Economic Analysis of Research Spillovers: Implications for the Advanced Technology Program*. NIST GCR 97-708. Gaithersburg, MD: National Institute of Standards and Technology.

Jaffe, A. B. 1998. "The Importance of 'Spillovers' in the Policy Mission of the ATP." *Journal of Technology Transfer* 23(1):11-19.

Jaffe, A. B. 2000. "The U.S. Patent System in Transition: Policy Innovation and the Innovation Process." *Research Policy* 29:531-557.

Jaffe, A. B., and J. Lerner. 2006. *Innovation and Its Discontents: How Our Broken Patent System Is Endangering Innovation and Progress, and What to Do About It.* Princeton, NJ: Princeton University Press.

Jaffe, A. B., J. Lerner, and S. Stern, eds. 2003. *Innovation Policy and the Economy: Volume 3.* Cambridge, MA: The MIT Press.

Jaffe, A. B., M. Trajtenberg, and R. Henderson. 1993. "Geographic Localization of Knowledge Spillovers as Evidenced by Patent Citations." *Quarterly Journal of Economics* 108(3):577-598.

Japan Patent Office. 2003. *Reports on Technology Trend and Patent Application: Life Science.* (in Japanese).

Japan Small Business Research Institute. 2002. *Survey of the Environment for Startups.* Tokyo, Japan: Japan Small Business Research Institute.

Jasanoff, Sheila, ed. 1997. *Comparative Science and Technology Policy.* Elgar Reference Collection. International Library of Comparative Pubic Policy, Volume 5. Cheltenham, UK: Edward Elgar.

Jorgenson, D. W. 2001. "Information Technology and the U.S. Economy." *American Economic Review* 91(1) March.

Jorgenson, D. W. 2001. "U.S. Economic Growth in the Information Age." *Issues in Science and Technology* Fall.

Jorgenson, Dale, and Kevin Stiroh. 2002. "Raising the Speed Limit: Economic Growth in the Information Age." In National Research Council. *Measuring and Sustaining the New Economy.* Dale Jorgenson and Charles Wessner, eds. Washington, DC: The National Academies Press.

Jorgenson, D. W., and K. Motohashi. 2003. "The Role of Information Technology in the Economy: Comparison between Japan and the United States." Prepared for *RIETI/KEIO Conference on Japanese Economy: Leading East Asia in the 21st Century?* Keio University. May 30.

Joy, William. 2000. "Why the Future Does Not Need Us." *Wired* 8.04. April.

Kaelble, Steve. 2004. "Good Neighbors: Indiana's Certified-Technology-park Program." *Indiana Business Magazine* July 1.

Kapur, Devesh. 2003. "Indian Diaspora as a Strategic Asset." *Economic and Political Weekly* 38(5):445-448.

Kapur, Surinder. 2006. "Nurture New Technology and Innovation, Stay Competitive." *The Financial Express* November 16.

The Kauffman Foundation and the Information Technology and Innovation Foundation. 2007. *The 2007 State New Economy Index. <http://www.kauffman.org/items.cfm?itemID=766>.*

Kawamoto, Takuji. 2004. "What Do the Purified Solow Residuals Tell Us about Japan's Lost Decade?" *Bank of Japan IMES Discussion Paper Series.* No. 2004-E-5. Tokyo, Japan: Bank of Japan.

Kenney, Martin, ed. 2000. *Understanding Silicon Valley: The Anatomy of an Entrepreneurial Region.* Stanford, CA: Stanford University Press.

Kim, Yong-June. 2006. "A Korean Perspective on China's Innovation System." Presentation at *Industrial Innovation in China.* Levin Institute Conference. July 24-26.

Kneller, R. 2003. "University-Industry Cooperation and Technology Transfer in Japan Compared with the U.S.: Another Reason for Japan's Economic Malaise?" *Journal of International Economic Law* 24:329-450.

Koizumi, Kei. 2007. "Historical Trends in Federal R&D." In *AAAS Report XXXII: Research and Development FY2008.* AAAS Publication Number 07-1A. Washington, DC: American Association for the Advancement of Science.

Kondo, Masayuki. 2003. "Chinese Model to Create High-Tech Start-Ups from Universities and Research Institutes." In M. von Zedtwitz, G. Haour, T. Khalil, and L. Lefebvre, eds. *Management of Technology: Growth through Business, Innovation and Entrepreneurship*. Oxford, UK: Pergamon Press.

Kondo, Masayuki. 2004. "Policy Innovation in Science and Technology in Japan—from S&T Policy to Innovation Policy." (In Japanese.) *Journal of Science Policy and Research Management* 19(3/4):132-140.

Kondo, Masayuki. 2004. "University spin-offs in Japan." *Asia Pacific Tech Monitor*. March-April 2004. Pp. 37-43. Asian and Pacific Centre for Transfer of Technology, ESCAP, UN.

Kondo, Masayuki. 2005. "Spin-offs from Public Research Institutes as Domestic Technology Transfer Means—The Case of RIKEN and Riken Industrial Group." (In Japanese.) *Development Engineering* 11:31-41.

Kondo, Masayuki. 2006. "University-Industry Partnerships in Japan." Presentation at the conference, *21st Century Innovation Systems for the United States and Japan: Lessons from a Decade of Change*. Tokyo, Japan. The National Academies, NISTEP of Japan, and The Institute of Innovation Research of Hitotsubashi University. *<http://www.nistep.go.jp/IC/ic060110/pdf/5-2.pdf>*.

Kortum, S., and J. Lerner. 1999. "What is Behind the Recent Surge in Patenting?" *Research Policy* 28:1-22.

Koschatzky, Knut. 2003. "The Regionalization of Innovation Policy: New Options for Regional Change?" In G. Fuchs and Phil Shapira, eds. *Rethinking Regional Innovation: Path Dependency or Regional Breakthrough?* London, UK: Kluwer.

Krugman, P. 1991. *Geography and Trade*, Cambridge, MA: MIT Press.

Krugman, Paul. 1994. "Competitiveness: A Dangerous Obsession." *Foreign Affairs* March/April.

Kuhlmann, Stephan and Jakob Edler. 2003. "Scenarios of Technology and Innovation Policies in Europe: Investigating Future Governance—Group of 3." *Technological Forecasting & Social Change* 70.

Kumar, Deepak. 1995. *Science and the Raj: 1857–1905*. New York: Oxford University Press.

Lall, Sanjaya. 2002. "Linking FDI and Technology Development for Capacity Building and Strategic Competitiveness." *Transnational Corporations* 11(3):39-88.

Lanjouw, J. O., and I. Cockburn. 2000. "Do Patents Matter? Empirical Evidence after GATT." NBER Working Paper No. 7495.

Lanjouw, J. O., and J. Lerner 1997. "The Enforcement of Intellectual Property Rights: A Survey of the Empirical Literature." NBER Working Paper No. W6296. Available at *<http://ssrn.com/abstract=226053>*.

Laredo, Philippe, and Philippe Mustar, eds. 2001. *Research and Innovation Policies in the New Global Economy: An International Perspective*. Cheltenham, UK: Edward Elgar.

Larosse, J. 2004. *Towards a 'Third Generation' Innovation Policy in Flanders: Policy Profile of the Flemish Innovation.* Brussels, Belgium: IWT.

Lazowska, Edward D., and David Paterson. 2005. "An Endless Frontier Postponed." *Science* 308(5723):757. May 6.

Lebra, Takie Sugiyama. 1971. "The Social Mechanism of Guilt and Shame: The Japanese Case." *Anthropological Quarterly* 44(4):241-255.

Lehman, Nicholas. 2001. "Dreaming About War." *The New Yorker* July 16. Available at *<http://www.comw.org/qdr/0107lemann.html>*.

Lembke, Johan. 2002. *Competition for Technological Leadership: EU Policy for High Technology.* Cheltenham, UK: Edward Elgar.

Lemley, Mark. 2000. "Rational Ignorance at the Patent Office." Berkeley Program in Law and Economics Working Paper Series No. 19.

Lemola, Tarmo. 2002. "Convergence of National Science and Technology Policies: The Case of Finland." *Research Policy* 31(8-9):1481-1490.

Lerner, J. 1995. "Patenting in the Shadow of Competitors." *Journal of Law and Economics* 38(Oct):463-495.

Lerner, Joshua. 1999. "Public Venture Capital." In National Research Council. *The Small Business Innovation Program: Challenges and Opportunities*. Charles W. Wessner, ed. Washington, DC: National Academy Press.

Lerner, J. 2002. "Patent Policy Shifts and Innovation Over 150 Years." *American Economic Review P&P.* (May).

Lerner, J., and J. Tirole. 2004. "Efficiency of Patent Pools." *American Economic Review* 94(3):691-711.

Leslie, Stuart, and Robert Kargon. 2006. "Exporting MIT." *Osiris* 21:110-130.

Levin, R. C., A. K. Klevorick, R. R. Nelson, and S. G. Winter. 1987. "Appropriating the Returns to Industrial R&D." *Brookings Papers on Economic Activity* 783-820.

Levitt, Rachelle, ed. 1987. *The University/Real Estate Connection: Research Parks and Other Ventures*. Washington, DC: Urban Land Institute.

Lewis, James A. 2005. *Waiting for Sputnik: Basic Research and Strategic Competition*. Washington, DC: Center for Strategic and International Studies.

Leyden, D. P., A. N. Link, and D. S. Siegel. 2008. "A Theoretical and Empirical Analysis of the Decision to Locate on a University Research Park." *IEEE Transactions on Engineering Management* 55(1):23-28.

Li, Jianjun. 2006. "The Development and Opening of Tianjin Binhai: New Area & China's Biotechnical Innovations." Presentation at *Industrial Innovation in China*. Levin Institute Conference. July 24-26.

Licklider, J. C. R. 1960. "Man-Computer Symbiosis." *IRE Transactions on Human Factors in Electronics*. March.

Lin, Otto. 1998. "Science and Technology Policy and its Influence on the Economic Development of Taiwan." In Henry S. Rowen, ed., *Behind East Asian Growth: The Political and Social Foundations of Prosperity*. New York: Routledge.

Lindh, T., and Ohlsson, H. 1996. "Self-Employment and Windfall Gains: Evidence from the Swedish Lottery." *Economic Journal* 106:1515-1526.

Lindelöf, P., and H. Löfsten. 2003. "Science Park Location and New Technology-Based Firms in Sweden: Implications for Strategy and Performance." *Small Business Economics* 20(3):245-258.

Lindelöf, P. and H. Löfsten. 2004. "Proximity as a Resource Base for Competitive Advantage: University-Industry Links for Technology Transfer." *Journal of Technology Transfer* 29(3-4): 311-326.

Link, A. N. 1981. "Basic Research and Productivity Increase in Manufacturing: Some Additional Evidence." *American Economic Review* 71(5):1111-1112.

Link, A. N. 1981. *Research and Development Activity in U.S. Manufacturing*. New York: Praeger.

Link, A. N. 1995. *A Generosity of Spirit: The Early History of the Research Triangle Park*. Research Triangle Park, NC: University of North Carolina Press for the Research Triangle Park Foundation.

Link, A. N. 2002. *From Seed to Harvest: The Growth of the Research Triangle Park*. Research Triangle Park, NC: University of North Carolina Press for the Research Triangle Park Foundation.

Link, A. N., and K. R. Link. 2003. "On the Growth of U.S. Science Parks." *Journal of Technology Transfer* 28(1):81-85.

Link, A. N., and J. T. Scott. 1998. *Public Accountability: Evaluating Technology-Based Institutions*. Norwell, MA: Kluwer Academic Publishers.

Link, A. N., and J. T. Scott. 2001. "Public/Private Partnerships: Stimulating Competition in a Dynamic Market." *International Journal of Industrial Organization* 19(5):763-794.

Link, A. N., and J. T. Scott. 2003. "The Growth of Research Triangle Park." *Small Business Economics* 20(2):167-175.

Link, A. N., and J. T. Scott. 2003. "U.S. Science Parks: The Diffusion of an Innovation and Its Effects on the Academic Mission of Universities." *International Journal of Industrial Organization* 21(9):1323-1356.

Link, A. N., and J. T. Scott. 2005. "Opening the Ivory Tower's Door: An Analysis of the Determinants of the Formation of U.S. University Spin-Off Companies." *Research Policy* 34(7):1106-1112.

Link, A. N., and J. T. Scott. 2006. "U.S. University Research Parks." *Journal of Productivity Analysis* 25(1):43-55.

Link, A. N., and J. T. Scott. 2007. "The Economics of University Research Parks." *Oxford Review of Economic Policy* 23(4):661-674.

Link, A. N., and D. S. Siegel. 2003. *Technological Change and Economic Performance*. London, UK: Routledge.

Litan, Robert E., and Lesa Mitchell. 2008. "Should Universities be Agents of Economic Development?" *Astra Briefs* 7(7-8).

Löfsten, Hans, and Peter Lindelöf. 2002. "Science Parks and the Growth of New Technology-based Firms—Academic-industry Links, Innovation and Markets." *Research Policy* 31(6):859-876.

Long, Guogiang. 2005. "China's Policies on FDI: Review and Evaluation." In Theodore H. Moran, Edward M. Graham, and Magnus Blomström, eds., *Does Foreign Direct Investment Promote Development?* Washington, DC: Institute for International Economics.

Luger, M. I. 2001. "Introduction: Information Technology and Regional Economic Development." *Journal of Comparative Policy Analysis: Research & Practice*.

Luger, M. I. 2007. "Smart Places for Smart People: Using Cluster-based Planning in the 21st Century." In *Creating Enterprise: Igniting Innovation through Business-University-Government Networks*. Cheltenham, UK: Edward Elgar.

Luger, M. I., and H. A. Goldstein. 1991. *Technology in the Garden*. Chapel Hill, NC: University of North Carolina Press.

Luger, M. I., and Harvey A. Goldstein. 2006. *Research Parks Redux: The Changing Landscape of the Garden*. Washington, DC: U.S. Economic Development Administration.

Lundstörm, A., and L. Stevenson. 2001. *Entrepreneurship Policy for the Future*. Stockholm, Sweden: FSF.

Lynn, Leonard. 2006. "Collaborative Advantage and China's Evolving Position in the Global Technology System." Presentation at *Industrial Innovation in China*. Levin Institute Conference. July 24-26.

Macher, Jeffrey, David Mowery, and David Hodges. 1999. "Semiconductors." In National Research Council. *U.S. Industry in 2000: Studies in Competitive Performance*. David C. Mowery, ed. Washington, DC: National Academy Press.

Maddison, Angus, and Donald Johnston. 2001. *The World Economy: A Millennial Perspective*. Paris, France: Organisation for Economic Co-operation and Development.

Malecki, E. J. 1991. *Technology and Economic Development*. New York: John Wiley.

Mandel, Michael. 2008. "A Better Way to Track the Economy: A Groundbreaking Commerce Dept. Report Could Lead to New Yardsticks for Measuring Growth." *BusinessWeek* (January 28):29.

Mani, Sunil. 2004. "Government, Innovation and Technology Policy: An International Comparative Analysis." *International Journal of Technology and Globalization* 1(1).

Mansfield, E. 1986. "Patents and Innovation: An Empirical Study." *Management Science* 32:173-181.

Mansfield, E. 1996. *Estimating Social and Private Returns from Innovations Based on the Advanced Technology Program: Problems and Opportunities*. GCR 99-780. Gaithersburg, MD: National Institute of Standards and Technology.

Mansfield, E. 1996. "How Fast Does New Industrial Technology Leak Out?" *Journal of Industrial Economics* 34(2):217-224.

Markoff, J. 2005. "Clouds Over 'Blue Sky' Research Agency." *New York Times* May 4. P. 12.

Markoff, J. 2005. "3 Technology Rivals Join to Finance Research at Berkeley." *New York Times* December 15. P. C5.

Marshall, Andrew. 1993. "Some Thoughts on Military Revolutions–Second Version," Memorandum for the Record. August 23.

Mazuzan, George. 1988. *The National Science Foundation: A Brief History (1950-1985)*. NSF 88-16. Arlington, VA: The National Science Foundation.

McKibben, William. 2003. *Enough: Staying Human in an Engineered Age*. New York: Henry Holt & Co.

Megginson, William L. 2004. "Towards a Global Model of Venture Capital?" *Journal of Applied and Corporate Finance* 16(1).

Merges, R. P. 1996. "A Comparative Look at Intellectual Property Rights and the Software Industry." In David C. Mowery, ed. *The International Software Industry*. New York: Oxford University Press.

Merges, R. P. 1999. "As Many as Six Impossible Patents before Breakfast: Property Rights for Business Concepts and Patent System Reform." *Berkeley Technology Law Journal.*

Merges, R., and R. R. Nelson. 1990. "On the Complex Economics of Patent Scope." *Columbia Law Review* 90:839-916.

Merton, R. K. 1968. "The Matthew Effect in Science." *Science* 159(3810):56-63.

METI (Small and Medium Enterprise Agency, Ministry of Economy, Trade and Industry). 2003. *The 2003 White Paper on Small and Medium Enterprises in Japan*. Tokyo: Japan Small Business Research Institute.

METI. 2005. *The 2005 White Paper on Small and Medium Enterprises in Japan*. Tokyo: Japan Small Business Research Institute.

MEXT. 2004. *Annual Report on Science and Technology Promotion Measures 2004: White Paper on Science and Technology*. (*Heisei 16 nendo Kagaku-Gijutu Sinkou-ni-kansuru Nenji-houkoku*). MEXT.

Meyer-Krahmer, Frieder. 2001. "The German Innovation System." Pp. 205-252 in P. Larédo and P. Mustar, eds. *Research and Innovation Policies in the New Global Economy: An International Comparative Analysis*. Cheltenham, UK: Edward Elgar.

Meyer-Krahmer, Frieder. 2001. "Industrial Innovation and Sustainability—Conflicts and Coherence." Pp. 177-195 in Daniele Archibugi and Bengt-Ake Lundvall, eds. *The Globalizing Learning Economy.* New York: Oxford University Press.

Middleton, Andrew, and Steven Bowns. With Keith Hartley and James Reid. 2006. *The Effect of Defense R&D on Military Equipment Quality, Defense and Peace Economics* 17(2):117-139. April.

Miller, Roger, and Marcel Cote. 1987. *Growing the Next Silicon Valley: A Guide for Successful Regional Planning*. Toronto: D. C. Heath and Company.

Ministry of Economy, Trade and Industry. 2006. *New Economic Growth Strategies 2006.* Tokyo, Japan: Research Institute of Economy, Trade and Industry. (In Japanese).

Ministry of Health, Labor and Welfare. 2002. *Visions of Pharmaceutical Industry in Japan.* <*http://www.mhlw.go.jp/houdou/2002/04/h0409-1.html*> (In Japanese).

Mitra, Raja. 2006. "India's Potential as a Global R&D Power." In Magnus Karlsson, ed. *The Internationalization of Corporate R&D*. Östersund: Swedish Institute for Growth Policy Studies.

Miyata, Shinpei. 1983. *A Free Paradise for Scientists.* (In Japanese.) Bungeishunju.

Mody, Ashok, and Carl Dahlman. 1992. "Performance and Potential of Information Technology: An International Perspective." *World Development* 20(12):1703-1719.

Moore, Gordon. 2003. "The SEMATECH Contribution." In National Research Council. *Securing the Future: Regional and National Programs to Support the Semiconductor Industry*. Charles W. Wessner, ed. Washington, DC: The National Academies Press.

Mooris, P. R. 1990. *A History of The World Semiconductor Industry.* Stevenage, UK: Peter Peregrinus Ltd.

Morrow, Daniel S. 2003. *Dr. J. Craig Venter–Oral History*. Computer World Honors Program. April 21. Available at <*http://cwheroes.org/archives/histories/venter*>.

Moser, P. 2001. "How Do Patent Laws Influence Innovation? Evidence from Nineteenth Century World Fairs." University of California at Berkeley: working paper.

Motohashi, Kazuyuki. 2005. "University–industry Collaborations in Japan: The Role of New Technology-based Firms in Transforming the National Innovation System." *Research Policy* 34:583-594.

Motohashi, Kazuyuki, and Xiao Yun. 2007. "China's Innovation System Reform and Growing Industry and Science Linkages." *Research Policy* 36:1251-1260.

Mowery, D. C., and B. N. Sampat. 2005. "The Bayh-Dole Act of 1980 and University-Industry Technology Transfer: A Model for Other OECD Governments?" *Journal of Technology Transfer* 30(1/2):115-127.

Mowery, D. C., and A. A. Ziedonis. 2002. "Academic Patent Quality and Quantity Before and After the Bayh-Dole Act in the United States." *Research Policy* 31(3):399-418.

Mowery, D., R. Nelson, B. Sampat, and A. Ziedonis. 2001. "The Growth of Patenting and Licensing by U.S. Universities: An Assessment of the Effects of the Bayh–Dole Act of 1980." *Research Policy* 30.

Mowery, D., R. Nelson, B. Sampat, and A. Ziedonis. 2003. *The Ivory Tower and Industrial Innovation*. Stanford, CA: Stanford University Press.

Murali, Kanta. 2003. "The IIT Story: Issues and Concerns." *Frontline* 20(3).

Murphy, L. M., and P. L. Edwards. 2003. *Bridging the Valley of Death: Transitioning from Public to Private Sector Financing*. Golden, CO: National Renewable Energy Laboratory. May.

Mustar, Phillipe, and Phillipe Laredo. 2002. "Innovation and Research Policy in France (1980–2000) or the Disappearance of the Colbertist State." *Research Policy* 31:55-72.

Nadiri I. 1993. *Innovations and Technological Spillovers*. NBER Working Paper No. 4423. Boston, MA: National Bureau of Economic Research.

Nagaoka Sadao. 2005. "How Does Priority Rule Work? Evidence from the Patent Examination Records in Japan." A Paper Presented for Patent Statistics and Innovation Research Workshop, November 25. Research Center for Advanced Science and Technology, University of Tokyo.

Nagaoka, Sadao. 2006. "Reform of Patent System in Japan and Challenges." Presentation at the conference, *21st Century Innovation Systems for the United States and Japan: Lessons from a Decade of Change*. Tokyo, Japan. The National Academies, NISTEP of Japan, and The Institute of Innovation Research of Hitotsubashi University. *<http://www.nistep.go.jp/IC/ic060110/pdf/5-2.pdf>*.

Nagaoka, Sadao. 2007. "Assessing the R&D Management of Firms by Patent Citation: Evidence from the U.S. Patents." *Journal of Economics & Management Strategy* 16(1).

Nagaoka, Sadao, Naotoshi Tsukada, and Tomoyuki Shimbo. 2006. "The Emergence and Structure of Essential Patents of Standards: Lessons from Three IT Standards." IIR Working Paper WP#06-08. Institute of Innovation Research, Hitotsubashi University.

Nail, John, and Hayden Brown. Forthcoming. *Classifying and Analyzing ATP Projects Using the North American Industry Classification System (NAICS)*. NISTIR 2005. Gaithersburg, MD: National Institute of Standards and Technology.

Nakamura, K., Y. Okada, and A. Tohei. 2006. "Does the Public Sector Make a Significant Contribution to Biomedical Research in Japan? A Detailed Analysis of Government and University Patenting, 1991-2002." Discussion Paper Series. Competition Policy Research Center. Fair Trade Commission of Japan. CPDP25-E. *<http://www.jftc.go.jp/cprc/DP/discussionpapers.html>*.

Nakayama, Yasuo, Mitsuaki Hosono, Nobuya Fukugawa, and Masayuki Kondo. 2005. "University-Industry Cooperation: Joint Research and Contract Research." (In Japanese.) *NISTEP Research Material*. No. 119. Tokyo, Japan: National Institute of Science and Technology Policy, Ministry of Education, Culture, Sports, Science and Technology.

Nanotechwire. 2005. "Governor Pataki Announces Historic Investments by IBM Global High-Tech Leaders in Nanoelectronics Manufacturing and Development." January 6. Accessed at *<http://www.nanotechwire.com/news.asp?nid=1453>*.

National Academy of Engineering. 2004. *The Engineer of 2020: Visions of Engineering in the New Century*. Washington, DC: The National Academies Press.

National Academy of Sciences/National Academy of Engineering/Insitute of Medicine. 2007. *Rising Above the Gathering Storm: Energizing and Employing America for a Brighter Economic Future*. Washington, DC: The National Academies Press.

National Institute of Science and Technology Policy and Mitsubishi Research Institute. 2005. *Government S&T Budget Analysis during the First and Second S&T Basic Plans.* (*Dai-ikki oyobi dai-niki Kagaku-Gijutu-Kihonkeikaku kikanchuu no kenkyuu-kaihatu-soushi no naiyou-bunseki*). NISTEP Report No. 84. March.

National Manufacturing Competitiveness Council. 2006. "The National Strategy for Manufacturing." New Delhi, India. March.

National Research Council. 1982. *Scientific Communication and National Security.* Washington, DC: National Academy Press.

National Research Council. 1987. *Balancing the National Interest: U.S. National Security Export Controls and Global Economic Competition*. Washington, DC: National Academy Press.

National Research Council. 1996. Conflict and Cooperation in National Competition for High-technology Industry. Washington, DC: National Academy Press.

National Research Council. 1999. *The Advanced Technology Program: Challenges and Opportunities.* Charles W. Wessner, ed. Washington, DC: National Academy Press.

National Research Council. 1999. *Funding a Revolution: Government Support for Computing Research*. Washington, DC: National Academy Press.

National Research Council. 1999. *Industry-Laboratory Partnerships: A Review of the Sandia Science and Technology Park Initiative.* Charles W. Wessner, ed. Washington, DC: National Academy Press.

National Research Council. 1999. *New Vistas in Transatlantic Science and Technology Cooperation.* Charles W. Wessner, ed. Washington, DC: National Academy Press.

National Research Council. 1999. *The Small Business Innovation Research Program: Challenges and Opportunities.* Charles W. Wessner, ed. Washington, DC: National Academy Press.

National Research Council. 2000. *The Small Business Innovation Research Program: A Review of the Department of Defense Fast Track Initiative.* Charles W. Wessner, ed. Washington, DC: National Academy Press.

National Research Council. 2000. *U.S. Industry in 2000: Studies in Competitive Performance*. David C. Mowery, ed. Washington, DC: National Academy Press.

National Research Council. 2001. *The Advanced Technology Program: Assessing Outcomes.* Charles W. Wessner, ed. Washington, DC: National Academy Press.

National Research Council. 2001. *Building a Workforce for the Information Economy.* Washington, DC: National Academy Press.

National Research Council. 2001. *Capitalizing on New Needs and New Opportunities: Government-Industry Partnerships in Biotechnology and Information Technologies*. Charles W. Wessner, ed. Washington, DC: National Academy Press.

National Research Council. 2001. *A Review of the New Initiatives at the NASA Ames Research Center.* Charles W. Wessner, ed. Washington, DC: National Academy Press.

National Research Council. 2001. *Trends in Federal Support of Research and Graduate Education.* Stephen A. Merrill, ed. Washington, DC: National Academy Press.

National Research Council. 2002. *Government-Industry Partnerships for the Development of New Technologies: Summary Report.* Charles W. Wessner, ed. Washington, DC: The National Academies Press.

National Research Council. 2003. *Government-Industry Partnerships for the Development of New Technologies: Summary Report.* Charles W. Wessner, ed. Washington, DC: The National Academies Press.

National Research Council. 2003. *Securing the Future: Regional and National Programs to Support the Semiconductor Industry*. Charles W. Wessner, ed. Washington, DC: The National Academies Press.

National Research Council. 2004. *Getting Up to Speed: The Future of Supercomputing*. Washington, DC: The National Academies Press.

National Research Council. 2004. *A Patent System for the 21st Century*. Washington, DC: The National Academies Press.

National Research Council. 2004. *Productivity and Cyclicality in Semiconductors: Trends, Implications, and Questions*. Dale W. Jorgenson and Charles W. Wessner, ed. Washington, DC: The National Academies Press.

National Research Council. 2004. *The Small Business Innovation Research Program: Program Diversity and Assessment Challenges*. Charles W. Wessner, ed. Washington, DC: The National Academies Press.

National Research Council. 2005. *Getting Up to Speed: The Future of Superconducting*. Susan L. Graham, Marc Snir, and Cynthia A. Patterson, eds. Washington, DC: The National Academies Press.

National Research Council. 2005. *Policy Implications of International Graduate Students and Postdoctoral Scholars in the United States*. Washington, DC: The National Academies Press.

National Research Council. 2007. *Enhancing Productivity Growth in the Information Age: Measuring and Sustaining the New Economy*. Dale W. Jorgenson and Charles W. Wessner, eds. Washington, DC: The National Academies Press.

National Research Council. 2007. *India's Changing Innovation System*. Charles W. Wessner and Sujai J. Shivakumar, eds. Washington, DC: The National Academies Press.

National Research Council. 2007. *Innovation Policies for the 21st Century*. Charles W. Wessner, ed. Washington, DC: The National Academies Press.

National Research Council. 2007. *SBIR and the Phase III Challenge of Commercialization*. Charles W. Wessner, ed. Washington, DC: The National Academies Press.

National Research Council. 2008. *An Assessment of the SBIR Program*. Charles W. Wessner, ed. Washington, DC: The National Academies Press.

National Research Council. 2008. *An Assessment of the SBIR Program at the Department of Energy*. Charles W. Wessner, ed. Washington, DC: The National Academies Press.

National Research Council. 2008. *An Assessment of the SBIR Program at the National Science Foundation*. Charles W. Wessner, ed. Washington, DC: The National Academies Press.

National Research Council. 2008. *Innovative Flanders: Innovation Policies for the 21st Century*. Charles W. Wessner, ed. Washington, DC: The National Academies Press.

National Research Council. 2009. *21st Century Innovation Systems for Japan and the United States: Lessons from a Decade of Change*. Sadao Nagaoka, Masayuki Kondo, Kenneth Flamm, and Charles Wessner, eds. Washington, DC: The National Academies Press.

National Research Council. 2009. *An Assessment of the SBIR Program at the National Institutes of Health*. Charles W. Wessner, ed. Washington, DC: The National Academies Press.

National Research Council. 2009. *An Assessment of the SBIR Program at the National Aeronautics and Space Administration*. Charles W. Wessner, ed. Washington, DC: The National Academies Press.

National Science Board. 2004. *Science and Engineering Indicators 2004*. Arlington, VA: National Science Foundation.

National Science Board. 2006. *Science and Engineering Indicators*, Arlington, VA: National Science Foundation.

National Science Board. 2008. *Science and Engineering Indicators 2008*. Arlington, VA: National Science Foundation.

National Science Foundation. 2004. *Science and Engineering Doctorate Awards: 2003*. NSF 05-300. Arlington, VA: National Science Foundation.

Needham, Joseph. 1954-1986. *Science and Civilization in China* (five volumes). Cambridge: Cambridge University Press.

Nelson, R. R. 1993. *National Innovation Systems: A Comparative Analysis*. New York: Oxford University Press.

Nelson, R. R., and K. Nelson. 2002. "Technology, Institutions, and Innovation Systems." *Research Policy* 31:265-272.

Nelson, R. R., and N. Rosenberg. 1993. "Technical Innovation and National Systems." In *National Innovation Systems: A Comparative Analysis*. Richard R. Nelson, ed. Oxford, UK: Oxford University Press.

Nelson, R. R., and S. G. Winter. 1982. *An Evolutionary Theory of Economic Change*. Cambridge, MA: Harvard University Press.

Nishimura, Kiyohiko, G., Takanobu Nakajima, and Kozo Kiyota. 2003. "Does the Natural Selection Mechanism Still Work in Severe Recessions? Examination of the Japanese Economy in the 1990s." *Journal of Economic Behavior and Organization* 58:53-78.

NISTEP (National Institute of Science and Technology Policy). 2004. *Study for Evaluating the Achievements of the S&T Basic Plans in Japan—FY2003 Highlights*. (In Japanese.) NISREP REPORT No. 74.

NISTEP. 2005. *Study for Evaluating the Achievements of the S&T Basic Plans in Japan—Highlights* (In Japanese.) NISTEP REPORT No. 83.

NISTEP. 2005. *Survey on the Performance of the 1st and the 2nd Science and Technology Basic Plan*. NISTEP Report No. 84. (In Japanese).

NISTEP and MRI. 2004. *Achievements and Issues of Major Policies for Industry-Academy-Government Collaboration and Regional Innovation—Study for FY2003*. (In Japanese.) NISTEP REPORT No. 78.

Odagari, H. 1999. "University-Industry Collaboration in Japan: Facts and Interpretations." In Lewis M. Branscomb, Fumio Kodama, and Richard Florida, eds. *Industrializing Knowledge: University-Industry Linkage in Japan and the United States*. Cambridge, MA: The MIT Press.

Odagiri, H., and A. Goto. 1993. "The Japanese System of Innovation: Past, Present and Future." In R. R. Nelson, ed. *National Systems of Innovation*. Oxford, UK: Oxford University Press.

Odagiri, H., and A. Goto. 1996. *Technology and Industrial Development in Japan: Building Capabilities by Learning, Innovation and Public Policy*. Oxford, UK: Oxford University Press.

Office of Pharmaceutical Industry Research. 2000. *Seiyaku Sangyou Ron* (*Analyses on Pharmaceutical Industry*). Japan Pharmaceutical Manufacturers Association. (In Japanese).

Ohkohchi Commemorative Foundation. 1954. *Masatoshi Ohkohchi*. (In Japanese.) Tokyo, Japan: Nikkan Kogyo Shimbunsha.

Organisation for Economic Co-operation and Development. 1999. *Boosting Innovation: The Cluster Approach*. Paris, France: Organisation for Economic Co-operation and Development.

Organisation for Economic Co-operation and Development. 1999. *Managing National Innovation Systems*. Paris, France: Organisation for Economic Co-operation and Development.

Organisation for Economic Co-operation and Development. 2001. *The New Economy: Beyond the Hype—The OECD Growth Project*. Paris, France: Organisation for Economic Co-operation and Development.

Organisation for Economic Co-operation and Development. 2001. *Social Sciences and Innovation*. Washington, DC: Organisation for Economic Co-operation and Development.

Organisation for Economic Co-operation and Development. 2004. "Summary Report: Micro-policies for Growth and Productivity." DSTI/IND(2004)7. Paris, France: Organisation for Economic Co-operation and Development. October.

Organisation for Economic Co-operation and Development. 2006. *Main Science and Technology Indicators*. Paris, France: Organisation for Economic Co-operation and Development.

Orszag, Peter, and Thomas Kane. 2003. "Funding Restrictions at Public Universities: Effects and Policy Implications." *Brookings Institution Working Paper*. September.

Oughton, Christine. 1997. "Competitiveness in the 1990s." *The Economic Journal* 107(444): 1486-1503.

Oughton, Christine, Mikel Landabaso, and Kevin Morgan. 2002. "The Regional Innovation Paradox: Innovation Policy and Industrial Policy." *The Journal of Technology Transfer* 27(1).

Owens, William, with Edward Offley. 2001. *Lifting the Fog of War*. Baltimore, MD: Johns Hopkins University Press.

Park, W. G., and J. C. Ginarte. 1997. "Intellectual Property Rights and Economic Growth." *Contemporary Economic Policy* XV(July):51-61.

Patel, P., and K. Pavitt. 1994. "National Innovation Systems: Why They are Important and How They Might Be Compared?" *Economic Change and Industrial Innovation.*

Peck, M. J., R. C. Levin, and Akira Goto. 1988. "Picking Losers: Public Policy Toward Declining Industries in Japan." In J. B. Shoven, ed. *Government Policy Toward Industry in the United States and Japan.* Cambridge, UK: Cambridge University Press. Pp. 165-239.

Peek, Joe, and Eric S. Rosengren. 2005. "Unnatural Selection: Perverse Incentives and the Misallocation of Credit in Japan." *The American Economic Review* 95(4):114-1166.

Pelsoci, Thomas. 2003. *Low-Cost Manufacturing Process Technology for Amorphous Silicon Detector Panels: Applications in Digital Mammography and Radiography*. NIST GCR 03-844. Gaithersburg, MD: National Institute of Standards and Technology.

Pelsoci, Thomas. 2005. *Photonic Technologies: Applications in Petroleum Refining, Building Controls, Emergency Medicine, and Industrial Materials Analysis.* NIST GCR 05-879. Gaithersburg, MD: National Institute of Standards and Technology.

Penrose, E. 1951. *The Economics of the International Patent System*. Baltimore, MD: Johns Hopkins University Press.

Perez, Carlota. 2002. *Technological Revolutions and Financial Capital*. Cheltenham, UK: Edward Elgar.

Phan, Phillip H., and Donald S. Siegel. 2006. "The Effectiveness of University Technology Transfer: Lessons Learned from Qualitative and Quantitative Research in the U.S. and U.K." Rensselaer Working Papers in Economics 0609. Troy, NY: Rensselaer Polytechnic Institute Department of Economics.

Phan, Phillip H., Donald S. Siegel, and Mike Wright. 2005. "Science Parks and Incubators: Observations, Synthesis and Future Research." *Journal of Business Venturing* 20(2):165-182.

Pharmaceuticals and Medical Device Agency. 2007. *Annual Report.* (In Japanese). *<http://www.pmda.go.jp/index.html>*.

Phillimore, J. 1999. "Beyond the Linear View of Innovation in Science Park Evaluation: An Analysis of Western Australian Technology Park." *Technovation* 19(11):673-680.

Polenske, Karen, Nicolas Rockler, et al. 2004. *Closing the Competitive Gap: A Retrospective Analysis of the ATP 2mm Project*. NIST GCR 03-856. Gaithersburg, MD: National Institute of Standards and Technology.

Popkin, Joel. 2003. *Inter-Industry Diffusion of Technology That Results from ATP Projects: Preliminary Research of the Potential Impacts of ATP Funding*. NIST GCR 03-848. Gaithersburg, MD: National Institute of Standards and Technology.

Porter, M. E., 1998. "Clusters and the New Economics of Competition." *Harvard Business Review* 76(6):77-90.

Porter, M. E., and M. Sakakibara. 2004. "Competition in Japan." *The Journal of Economic Perspectives* 18(1).

Posen, Adam. 1998. *Restoring Japan's Economic Growth*. Washington, DC: Peterson Institute for International Economics.

Posen, Adam S. 2001. "Japan." In Benn Steil, David G. Victor, and Richard R. Nelson, eds. *Technological Innovation and Economic Performance*. Princeton, NJ: Princeton University Press.

Powell, Jeanne, and Francisco Moris. 2002. *Different Timelines for Different Technologies*. NISTIR 6917. Gaithersburg, MD: National Institute of Standards and Technology.

President's Council of Advisors on Science and Technology. 2004. "Sustaining the Nation's Innovation Ecosystems." Washington, DC: Executive Office of the President. January.

President's Council of Advisors on Science and Technology. 2004. "Sustaining the Nation's Innovation System: Report on Information Technology Manufacturing and Competitiveness." Washington, DC: Executive Office of the President. January.

President's Information Technology Advisory Committee. 2005. "Cybersecurity: A Crisis of Prioritization." Report to the President. February.

PricewaterhouseCoopers. 2006. "China's Impact on the Semiconductor Industry: 2005 Update." New York: PricewaterhouseCoopers.

Procassini, Andrew. 1995. *Competitors in Alliance: Industry Associations, Global Rivalries, and Business-Government Relations*. New York: Greenwood Publishing.

Purvis, G. 2002. "Moving into the Real World." *Electronic Business*. July 1.

Raduchel, William. 2006. "The End of Stovepiping." In National Research Council. *The Telecommunications Challenge: Changing Technologies and Evolving Policies*, Charles W. Wessner, ed. Washington, DC: The National Academies Press.

Rai, Saritha. 2006. "India Becoming a Crucial Cog in the Machine at I.B.M." *The New York Times* June 5.

Rausser, Gordon C. 2000. Letter to the Editor. *Atlantic Monthly*. May 19. Accessed at <*http://www.cnr.berkeley.edu/pdf/dean_rausser/Atl_ltr_edt_5_2000.pdf*>.

Reid, T. R. 2004. *The United States of Europe: The New Superpower and the End of American Supremacy*. New York: Penguin Press.

Reiko, Yamada. 2001. "University Reform in the Post-massification Era in Japan: Analysis of Government Education Policy for the 21st Century." *Higher Education Policy* 14(4).

Reuters. 2006. "China Sees No Quick End to Economic Boom." February 21.

Rich, Ben. 1996. *Skunkworks*. New York: Back Bay Books.

Rodrik, Dani and Arvind Subramanian. 2004. "From 'Hindu Growth' to Productivity Surge: The Mystery of the Indian Growth Transition." NBER Working Paper 10376.

Roland, Alex. *Model Research: The National Advisory Committee for Aeronautics, 1915-1958*. Washington, DC: National Aeronautics and Space Administration. Available at <*http://history.nasa.gov/SP-4103/*>.

Romanainen, Jari. 2001. "The Cluster Approach in Finnish Technology Policy." Pp. 377-388 in Edward M. Bergman, Pim den Hertog, and David Charles, eds. *Innovative Clusters: Drivers of National Innovation Systems*. OECD Proceedings. Washington, DC: Organisation for Economic Co-operation and Development.

Romer, Paul. 1990. "Endogenous Technological Change." *Journal of Political Economy* 98:72-102.

Rosenberg, N., and R. Nelson. 1994. "American Universities and Technical Advance in Industry." *Research Policy* 23:325-348.

Rosenzweig, R., and B. Turlington. 1982. *The Research Universities and Their Patrons*. Berkeley, CA: University of California Press.

Rothaermel, F. T., and M. C. Thursby. 2005. "Incubator Firm Failure or Graduation? The Role of University Linkages." *Research Policy* 34(7):1076-1090.

Rothaermel, F. T., and M. C. Thursby. 2005. "University-Incubator Firm Knowledge Flows: Assessing Their Impact on Incubator Firm Performance." *Research Policy* 34(3):302-320.

Rowen, Henry S., and A. Maria Toyoda. 2002. "From Kiretsu to Start-ups: Japan's Push for High Tech Entrepreneurship." Asia-Pacific Research Center Working Paper. Stanford, CA.

Ruegg, Rosalie, and Irwin Feller. 2003. *A Toolkit for Evaluating Public R&D Investment: Models, Methods, and Findings from ATP's First Decade*. NIST GCR 03-857. Gaithersburg, MD: National Institute of Standards and Technology.

Ruttan, Vernon. 2002. *Technology, Growth and Development: An Induced Innovation Perspective*. Oxford, UK: Oxford University Press.

Ruttan, Vernon W. 2006. *Is War Necessary for Economic Growth, Military Procurement and Technology Development*. New York: Oxford University Press.

Ruttan, Vernon W. 2006. "Will Government Programs Spur the Next Breakthrough?" *Issues in Science and Technology* Winter.

Rutten, Roel, and Frans Boekema. 2005. "Innovation, Policy and Economic Growth: Theory and Cases." *European Planning Studies* 13(8).

Rycroft, Robert W., and Don E. Kash. 1999. "Innovation Policy for Complex Technologies." *Issues in Science and Technology* Fall.

Sakakibara, M., and L. Branstetter. 2001. "Do Stronger Patents Induce More Innovation? Evidence from the 1988 Japanese Patent law Reforms." *RAND Journal of Economics* 32:77-100.

Sakakibara, M., and L. Branstetter. 2002. *Measuring the Impact of ATP-Funded Research Consortia on Research Productivity of Participating Firms, A Framework Using Both U.S. and Japanese Data*. NIST GCR 02-830. Gaithersburg, MD: National Institute of Standards and Technology.

Sakamoto, Kozo, and Masayuki Kondo. 2004. "The Analysis of University-Industry Research Collaborations by Time Series and Corporate Characteristics." (In Japanese.) *Development Engineering* 10:11-26.

Saito, Ken. 1987. *Research on a New Concern: RIKEN Industrial Group*. (In Japanese.) Jichosha.

Sapolsky, Harvey M. 1990. *Science and the Navy—The History of the Office of Naval Research*. Princeton, NJ: Princeton University Press.

Saxenian, AnnaLee. 1994. *Regional Advantage: Culture and Competition in Silicon Valley and Route 128*. Cambridge, MA: Harvard University Press.

Schaffer, Teresita. 2002. "Building a New Partnership with India." *Washington Quarterly* 25(2):31-44.

Scherer, F. M. 2001. "U.S. Government Programs to Advance Technology." *Revue d'Economie Industrielle* 0(94):69-88.

Science Committee of the Council for Science and Technology of the Ministry of Education, Culture, Sports, Science and Technology. 2005. *Sciences Policy to Support Diversity in Research*. (*Kenkyuu-no-tayousei wo-sasaeru Gakujutu-seisaku*). October 13.

Scotchmer, Suzanne. 1996. "Protecting Early Innovators: Should Second-Generation Products Be Patentable?" *RAND Journal of Economics* 27:322-331.

Scotchmer, Suzanne. 2004. *Innovation and Incentives*. Cambridge, MA: The MIT Press.

Semiconductor Fabtech. 2005. "IBM & AMD Aim Alliance at the 22nm Frontier." November 1.

Semiconductor Fabtech. 2006. "Chinese Government Builds 300mm Fab." June 28.

Semiconductor Fabtech. 2006. "IBM and Partners Ready 45nm Low Power Process." August 30.

Semiconductor Fabtech. 2006. "IBM, Sony, Toshiba Take Technology Alliance Beyond 32nm." December 1.

Semiconductor Manufacturing International Corp. 2006. Form 20-F for Period Ending December 31, 2006.

Semiconductor Manufacturing International. 2006. "Q1 2006 Earnings Conference Call." April 27.

Shane, S. 2004. *Academic Entrepreneurship*. Cheltenham, UK: Edward Elgar Publishing.

Shang, Yong. 2006. "Innovation: New National Strategy of China." Presentation at *Industrial Innovation in China*. Levin Institute Conference. July 24-26.

Shanghai Municipal Government. 2004. *Notice of the Shanghai Municipal Government Regarding Distributing the Outline of Shanghai's Intellectual Property Strategy (2004-2010)*. September 14.

Shapiro, Carl. 2001. "Navigating the Patent Thicket: Cross License, Patent Pools and Standard-Setting." In Adam Jaffe, Joshua Lerner, and Scott Stern, eds. *Innovation Policy and the Economy*. Cambridge, MA: The MIT Press.

Shearmur, R., and D. Doloreux. 2000. "Science Parks: Actors or Reactors? Canadian Science Parks in their Urban Context." *Environment and Planning* 32(6):1065-1082.

Sheehan, Jerry, and Andrew Wyckoff. 2003. "Targeting R&D: Economic and Policy Implications of Increasing R&D Spending." DSTI/DOC(2003)8. Paris, France: Organisation for Economic Co-operation and Development.

Shenzhen Daily. 2005. "Nation May Introduce Antimonopoly Law." December 30.

Sherwin, Martin, and Kai Bird. 2005. *American Prometheus: The Triumph and Tragedy of J. Robert Openheimer*. New York: Alfred A. Knopf.

Shin, Roy W. 1997. "Interactions of Science and Technology Policies in Creating a Competitive Industry: Korea's Electronics Industry." *Global Economic Review* 26(4):3-19.

Shipp, Stephanie, Andrew Wang, Steve Campbell, Lorel Wisniewski, Kerry Levin, Jennifer O'Brien, and Westat. 2005. "Measuring Behavior Additionality in ATP Joint Venture Projects: Findings from the U.S. Advanced Technology Program." Paper presented at OECD Workshop on Behavioral Additionality, January 31-February 1, 2005. Vienna, Austria.

Siegel, D. S., P. Westhead, and M. Wright. 2003. "Assessing the Impact of Science Parks on Research Productivity: Exploratory Firm-Level Evidence from the United Kingdom." *International Journal of Industrial Organization* 21(9):1357-1369.

Siegel, Donald, David Waldman, and Albert Link. 2004. "Toward a Model of the Effective Transfer of Scientific Knowledge from Academicians to Practitioners: Qualitative Evidence from the Commercialization of University Technologies." *Journal of Engineering and Technology Management* 21(1-2):115-142.

Singh, Manmohan. 2006. Remarks by the Chairman of the Planning Commission on the release of the 11th Plan Approach Paper. Accessed at <*http://pmindia.nic.in/speech/content.asp?id=431*> October 18.

Skolnikoff, Eugene B. 1993. "Knowledge Without Borders? Internationalization of the Research Universities." *Daedalus* 122(4).

Small and Medium Enterprise Agency in Japan. 2002. "White Paper on Small and Medium Enterprises in Japan." Tokyo, Japan: Japan Small Business Research Institute.

Small and Medium Enterprise Agency in Japan. 2003. "White Paper on Small and Medium Enterprises in Japan." Tokyo, Japan: Japan Small Business Research Institute.

Smits, Ruud, and Stefan Kuhlmann. 2004. "The Rise of Systemic Instruments in Innovation Policy." *International Journal of Foresight and Innovation Policy* 1(1/2).

Soete, Luc G., and Bastiann J. ter Weel. 1999. "Innovation, Knowledge Creation and Technology Policy: The Case of the Netherlands." *De Economist* 147(3). September.

Sofouli, E., and N. S. Vonortas. 2007. "S&T Parks and Business Incubators in Middle-Sized Countries: The Case of Greece." *Journal of Technology Transfer* 32(5):525-544.

Solomon Associates. 1993. *The Advanced Technology Program: An Assessment of Short-Term Impacts: First Competition Participants*. Gaithersburg, MD: National Institute of Standards and Technology.

Solow, Robert. 2000. *Growth Theory: An Exposition*. New York: Oxford University Press. 2nd edition.

Solow, Robert. 2000. "Toward a Macroeconomics of the Medium Run." *Journal of Economic Perspectives* Winter.

Spence, Michael. 1974. *Market Signaling: Informational Transfer in Hiring and Related Processes*. Cambridge, MA: Harvard University Press.

Spencer, William, and T. E. Seidel. 2004. "International Technology Roadmaps: The U.S. Semiconductor Experience." In National Research Council. *Productivity and Cyclicality in Semiconductors: Trends, Implications, and Questions*. Dale W. Jorgenson and Charles W. Wessner, eds. Washington, DC: The National Academies Press.

Spencer, W. J., L. Wilson, and R. Doering. 2004. "The Semiconductor Technology Roadmap." *Future Fab International* 18.

Stanford University. 1999. *Inventions, Patents and Licensing: Research Policy Handbook*. Document 5.1. July 15.

State Council of the Peoples' Republic of China. 2006. *Outline of the National Medium- and Long-term Program on Scientific and Technological Development (2006-2020)*. February 9.

Statistics Bureau, Ministry of Internal Affairs and Communications. 2005. *Report on the Survey of Research and Development 2004*. Ministry of Internal Affairs and Communications.

Stephan, P. 2001. "Educational Implication of University-Industry Technology Transfer." *Journal of Technology Transfer* 26:199-205.

Sternberg, R. 1990. "The Impact of Innovation Centres on Small Technology-based Firms: The Example of the Federal Republic of Germany." *Small Business Economics* 2(2):105-118.

Stiglitz, Joseph. 2005. "The Ethical Economist." *Foreign Affairs*. Council on Foreign Affairs. November/December. Accessed at <*http:www.foreignaffairs.org*>.

Stokes, Donald E. 1997. *Pasteur's Quadrant: Basic Science and Technological Innovation*. Washington, DC: Brookings Institution Press.

Strout, E. 2005. "Gift of a Book Was a Key to Intel Founder's Big Donation to City College of New York." *Chronicle of Higher Education*. December 2. P. A27.

Su, Yun-Shan, and Ling-Chun Hung. 2008. "Spontaneous vs. Policy-driven: The Origin and Evolution of the Biotechnology Cluster." *Technological Forecast and Social Change*.

Swann, G. M. P. 1998. "Towards a Model of Clustering in High-Technology Industries." In G. M. P. Swann, M. Prevezer, and D. Stout, eds. *The Dynamics of Industrial Clustering*. Oxford, UK: Oxford University Press.

Talele, Chitram J. 2003. "Science and Technology Policy in Germany, India and Pakistan." *Indian Journal of Economics and Business* 2(1):87-100.

Tanaka, Nobua. 2005. Presentation at the International Forum on Technology Foresight and National Innovation Strategies. Seoul, Republic of Korea. November 4.

Tansley, A. G. 1939. "British Ecology During the Past Quarter Century: The Plant Community and the Ecosystem." *Journal of Ecology* 27(2):513-530.

Tassey, G. 1997. *The Economics of R&D Policy*. Westport, CT: Quorum Books.

Tassey, G. 2004. "Policy Issues for R&D Investment in a Knowledge-based Economy." *Journal of Technology Transfer* 29:153-185.

Tassey, G. 2007. *The Innovation Imperative*. Cheltenham, UK: Edward Elgar.

Teubal, Morris. 2002. "What Is the Systems Perspective to Innovation and Technology Policy and How Can We Apply It to Developing and Newly Industrialized Economies?" *Journal of Evolutionary Economics* 12(1-2).

Thelin, J. 2004. *A History of American Higher Education*. Baltimore, MD: Johns Hopkins University Press.

Thomas, J. R., and W. H. Schacht. 2008. "Patent Reform in the 110th Congress: Innovation Issues." Report for Congress. Washington, DC: Congressional Research Service. Order Code RL33996.

Thursby, J., and M. Thursby. 2002. "Who is Selling the Ivory Tower? Sources of Growth in University Licensing." *Management Science* 48:90-104.

Thursby, J., and M. Thursby. 2004. "Industry Perspectives on Licensing University Technologies: Sources and Problems." *AUTM Journal* P. 2000.

TMCNET. 2006. "Chip Giant Quickens Steps of Expansion." June 28.

U.S.–China Joint Commission on Commerce and Trade (JCCT). 2005. *Outcomes on Major U.S. Trade Concerns*. Washington, DC: The Office of the United States Trade Representative.

U.S. Congress. 1988. Omnibus Trade and Competitiveness Act of 1988 (P.L. 100-418, codified in 15 U.S.C. 278n.) and later amended by the American Technology Preeminence Act of 1991 (P.L. 102-245, codified in 15 U.S.C. 3701).

U.S. Congress. 2005. House Science Committee Hearing on the Future of Computer Science Research in the U.S. May 12. Testimony by Wm. A. Wulf, Pres., National Academy of Engineering, Prof. Thomas F. Leighton, Chief Scientist Akamai Tech. Inc., Joint Statement of the Computing Research Community, and Letters in Response to Committee Questions from Wm. Wulf and T. Leighton. July 2005.

U.S. Department of Commerce. 2008. *Innovation Measurement: Tracking the State of Innovation in the American Economy*, Report to the Secretary. Washington, DC: U.S. Department of Commerce. January. Available at *<http://www.innovationmetrics.gov/Innovation%20Measurement%2001-08.pdf>*.

U.S. Department of Energy. 2006. Press Release. "Department Requests $4.1 Billion Investment as Part of the American Competitiveness Initiative: Funding to Support Basic Scientific Research." February 2.

U.S. General Accounting Office. 2002. *Export Controls: Rapid Advances in China's Semiconductor Industry Underscore Need for Fundamental U.S. Policy Review*. GAO-020620. Washington, DC: U.S. General Accounting Office. April.

U.S. Small Business Administration. 2004. "Small Business by the Numbers," Office of Advocacy. Washington, DC: U.S. Small Business Administration. June.

Vaidyanathan, G. 2008. "Technology Parks in a Developing Country: The Case of India." *Journal of Technology Transfer* 33(3):285-299.

Van Atta, Richard. 2004. "Energy and Climate Change Research and the DARPA Model." Presentation to the Washington Roundtable on Science and Public Policy. November 3.

Van Atta, Richard. 2008. "Fifty Years of Innovation and Discovery." In *DARPA: 50 Years of Bridging the Gap*. Arlington, VA: Defense Advanced Research Projects Agency. April.

Van Atta, Richard, et al. 1991. *DARPA Technical Accomplishments*. Volumes I-V. Alexandria, VA: Institute for Defense Analysis.

Van Atta, Richard, et al. 1991. *DARPA Technological Accomplishments: An Historical Review of Selected DARPA Projects*. Alexandria, VA: Institute for Defense Analysis.

Van Atta, Richard, and Michael Lippitz. 2003. *Transformation and Transition: DARPA's Role in Fostering an Emerging Revolution in Military Affairs*. Volume 1: Overall Assessment. Alexandria, VA: Institute for Defense Analysis.

Van Looy, Bart, Marina Ranga, Julie Callaert, Koenraad Debackere, and Edwin Zimmermann. 2004. "Combining Entrepreneurial and Scientific Performance in Academia: Towards a Compounded and Reciprocal Matthew-effect?" *Research Policy* 33(3):425-441.

Van Looy, Bart, K. Debackere, and T. Magerman. 2005. *Assessing Academic Patent Activity: The Case of Flanders*. Leuven, Belgium: SOOS.

Vedovello, C. 1997. "Science Parks and University-Industry Interaction: Geographical Proximity between the Agents as a Driving Force." *Technovation* 17(9):491-502.

Vervliet, Greta. 2006. *Science, Technology, and Innovation*. Brussels, Belgium: Ministry of Flanders, Science and Innovation Administration.

Vest, C. 2005. "Industry, Philanthropy and Universities-The Roles and Influences of the Private Sector in Higher Education." 2005 Clark Kerr Lecture. University of California at Berkeley. September 13.

Veugelers, Reinhilde, Jan Larosse, Michele Cincera, Donald Carchon, and Roger Kalenga-Mpala. 2004. "R&D Activities of the Business Sector in Flanders: Results of the R&D Surveys in the Context of the 3% Target." Brussels, Belgium: IWT-Studies (46).

Wade, Robert. 1985. "The Market for Public Office: Why the Indian State Is Not Better at Development." *World Development* 13:467-497.

Wadhwa, Vivek, and Gary Gereffi. 2005. "Framing the Engineering Outsourcing Debate." Durham, NC: Duke University.

Walcott, Susan M. 2003. *Chinese Science and Technology Industrial Parks*. Aldershot, UK: Ashgate Publishing.

Waldrop, M. Mitchell. 1992. *Complexity: The Emerging Science at the Edge of Order and Chaos*. New York: Simon & Schuster.

Waldrop, M. Mitchell. 2001. *The Dream Machine*. New York: Viking Press.

Walsh, J., A. Arora, and W. Cohen. 2003. "Research Tool Patent and Licensing and Biomedical Innovation." In *Patents in the Knowledge-Based Economy*, W. Cohen and S. Merrill, eds. Washington, DC: The National Academies Press.

Walsh, J. P., and W. M. Cohen. 2004. "Does the Golden Goose Travel? A Comparative Analysis of the Influence of Public Research on Industrial R&D in the U.S. and Japan." Mimeo.

Wang, Changyong. 2005. "IPR Sails Against Current Stream." *Caijing Magazine*, October 17. Available online at *<http://caijing.hexun.com>*.

Wang, Liwei. 1993. "The Chinese Traditions Inimical to the Patent Law." *Northwestern Journal of International Law and Commerce* Fall.

Warsh, David. 2006. *Knowledge and the Wealth of Nations*. New York: W. W. Norton.

The Washington Post. 2006. "Chinese to Develop Sciences, Technology." February 10. P. A16.

Wen, Jiang, and Shinichi Kobayashi. 2001. "Exploring Collaborative R&D Network: Some New Evidence in Japan." *Research Policy* 30:1309-1319.

Wessner, Charles W. 2005. "Entrepreneurship and the Innovation Ecosystem." In David B. Audretsch, Heike Grimm, and Charles W. Wessner, eds. *Local Heroes in the Global Village: Globalization and the New Entrepreneurship Policies*. New York: Springer.

Westhead, P. 1995. "New Owner-Managed Businesses in Rural and Urban Areas in Great Britian: A Matched Pairs Comparison." *Regional Studies* 29(4):367-380.

Westhead, P. 1997. "R&D 'Inputs' and 'Outputs' of Technology-based firms Located On and Off Science Parks." *R&D Management* 27(1):45-61.

Westhead, P., and S. Batstone. 1998. "Independent Technology-based Firms: The Perceived Benefits of a Science Park Location." *Urban Studies* 35(12):2197-2219.

Westhead, P., and M. Cowling. 1995. "Employment Change in Independent Owner-Managed High-Technology Firms in Great Britain." *Small Business Economics* 7(2):111-140.

Westhead, P., and D. Storey. 1994. *An Assessment of Firms Located On and Off Science Parks in the United Kingdom*. London, UK: HMSO.

Westhead, P., and D. Storey. 1997. "Financial Constraints on the Growth of High-Technology Small Firms in the U.K." *Applied Financial Economics* 7(2):197-201.

Westhead, P., D. J. Storey, and M. Cowling. 1995. "An Exploratory Analysis of the Factors Associated with the Survival of Independent High-Technology Firms in Great Britain." In F. Chittenden, M. Robertson, and I. Marshall, eds. *Small Firms: Partnerships for Growth*. London, UK: Paul Chapman.

White House. 2006. "Fact Sheet: United States and India: Strategic Partnership." Press Release. March 2.

Wiener, Norbert. 1948. *Cybernetics or Control and Communication in the Animal and the Machine*. Cambridge, MA: The MIT Press.

World Bank. 2004. *Innovation Systems: World Bank Support of Science and Technology Development*. Vinod Kumar Goel, ed. Washington, DC: World Bank.

World Bank. 2006. *The Environment for Innovation in India*. South Asia Private Sector Development and Finance Unit. Washington, DC: World Bank.

World Bank. 2006. *World Development Indicators*. Accessed at *<http://devdata.worldbank.org/wdi2006/contents/cover.htm>*.

World Bank International Finance Corporation. 2006. *Doing Business in 2006: Creating Jobs*. Washington, DC: International Bank for Reconstruction and Development.

World Trade Organization Working Party on the Accession of China. 2001. *Report of the Working Party on the Accession of China*. WT/MN(01)/3. November 10.

Wu, Ching. 2006. *China to Build 30 New Science and Technology Parks*. SciDev.net. April 19.

Yan, Dai. 2004. "Anti-Monopoly Legislation on the Way." *China Daily* June 18.

Yang, Lei, ed. 2006. "Chinese WAPI Delegation Quits Prague Meeting." *Xinhua*. June 8. Available at *<http://news.xinhuanet.com>*.

Yasuda, T. 2005. "*Seisakukinyuu no riyou*" (The Utilization of Public Finance). In *Nippon no shinki kaigyou kigyou* (Startup Enterprises in Japan). K. Kustuna and T. Yasuda, ed. Hakutousha.

Yoshida, Fujio. 1967. "Preparation of Legal System for Capital Liberalization (Part 3)." Panel Discussion in *Zaikei Shoho*. July 17.

Yoshida, Junko. 2006. "Grenoble Lure: Un-French R&D." *EE Times* June 12.
Young, John. 2007. Info Memo for Secretary of Defense Robert M. Gates. DoD Science and Technology Program. August 24.
Yukio, Sato. 2001. "The Structure and Perspective of Science and Technology Policy in Japan." In Phillipe Laredo and Phillipe Mustar, eds. *Research and Innovation Policies in the New Global Economy: An International Comparative Analysis*. Cheltenham, UK: Edward Elgar.
Zachary, G. Pascal. 1999. *Endless Frontier: Vannevar Bush, Engineer of the American Century*. Cambridge, MA: The MIT Press.
Zeigler, Nicholas J. 1997. *Governing Ideas: Strategies for Innovation in France and Germany*. Ithaca, NY: Cornell University Press.
Zemin, Jiang, General Secretary of the Communist Party of China Central Committee. 1999. Keynote Speech at the National Technological Innovation Conference, August 23.
Ziedonis, R. H., and B. H. Hall. 2001. "The Effects of Strengthening Patent Rights on Firms Engaged in Cumulative Innovation: Insights from the Semiconductor Industry." In Gary Libecap, ed., *Entrepreneurial Inputs and Outcomes: New Studies of Entrepreneurship in the United States*. Vol. 13 of *Advances in the Study of Entrepreneurship, Innovation, and Economic Growth*. Amsterdam, The Netherlands: Elsevier Science.
Zue, Victor. 2008. "Introduction to CSAIL." MIT. April 15.